Learning and Applying SolidWorks 2011-2012 Step-by-Step

L. Scott Hansen, Ph.D.

Department Chair of Engineering Technology and
Construction Management and Associate Professor
of Engineering Technology

College of Science and Engineering

Southern Utah University
Cedar City, Utah

Industrial Press, Inc.
New York

Copyright © 2011 by Industrial Press Inc., New York.

ISBN 9780831134433
Sixth Edition, August 2011

Sponsoring Editor: John Carleo
Cover Design: Janet Romano

Printed in the United States of America. All rights reserved. This book, or any parts thereof, may not be reproduced, stored in a retrieval system, or transmitted in any form without the permission of the publisher.

Limits of Liability and disclaimer of Warranty
The author and publisher make no claim of warranty, written, expressed, assumed or implied.

Industrial Press Inc.
989 Avenue of the Americas, 19th Floor
New York, NY 10018

10 9 8 7 6 5 4 3 2 1

Note to the Reader

This book provides clear and concise applied instruction in order to help you develop a mastery of *SolidWorks*. Almost every instruction includes a graphic illustration to aid in clarifying that instruction. Software commands appear in **bold** or in "quotation marks" for anyone who prefers not to read every word of the text. Most illustrations also include small pointer arrows and text to further clarify instructions.

This book was written for classroom instruction or self-study, including for individuals with no solid modeling experience at all. You will begin at a very basic level, but by the time you finish you will be completing complex functions.

For any organization requiring additional help, I am available for onsite training. Please contact me at hansens@suu.edu

Scott Hansen
Cedar City, Utah

Table of Contents

Chapter 1 Getting Started 1

    Create a simple sketch using the Sketch commands 10
    Dimension a sketch using the Smart Dimension command. 18
    Extrude a sketch using the Features commands. 44
    Create a fillet using the Fillet command. 46
    Create a hole using the Extruded Cut command. 59
    Create a counter bore using the Hole Wizard command 65

Chapter 2 Learning More Basics. 77

    Revolve a sketch using the Revolve Boss/Base command. 94
    Create a hole using the Extruded Cut command. 113
    Create a series of holes using the Circular Pattern command. 118

Chapter 3 Learning To Detail Part Drawings. 127

    Create an Orthographic View using the Drawing commands 167

Chapter 4 Advanced Detail Drawing Procedures. 179

    Create an Auxiliary View using the Drawing commands 182
    Create a Section View using the Drawing commands 186
    Dimension views using the Drawing commands. 188
    Create Text using the Drawing Annotation commands 192

Chapter 5 Learning To Edit Existing Solid Models 201

    Edit the part using the Edit Sketch command 229
    Edit the part using the Edit Feature command 237

Chapter 6 Advanced Design Procedures. 259

    Learn to use the Front, Top, and Right Planes 267
    Learn to use the Wireframe viewing command 269
    Learn to Convert Entities on to a new sketch 271
    Learn to use the Shell command. 284

Chapter 7 Introduction To Assembly View Procedures 327

    Learn to insert existing solid models using the Assembly commands 328
    Learn to constrain all parts using the Mate command. 340
    Learn to edit/modify parts while in the Assembly command 367
    Learn to simulate motion using the Motion Study command. 375

Chapter 8 Introduction to Advanced Commands. 381

    Learn to use the Swept Boss/Base command. 387
    Learn to use the use the Plane command. 399
    Learn to use the Lofted Boss/Base command. 408

Chapter 9 Introduction to SimulationXpress 413

    Learn to run an analysis on a part. 415
    Learn to interpret and analysis on a part. 429

Chapter 10 Introduction to Mechanical Mates. 435

    Learn to use the Cam mechanical mate. 447

Chapter 11 Advanced Work Plane Procedures 455

    Learn to create points on a solid model 460
    Learn to create an offset work plane 467
    Learn to create an offset Extruded cut 472

Chapter 12 Introduction to Creating a Helix 477

    Learn how to create a Helix 478

Chapter 13 Introduction to Importing .DWG files 487

    Learn to create a simple .DWG file 488
    Learn to import a .DWG file into a 1 view drawing 489
    Learn to import a .DWG file into the solid model area and create a solid model . . . 496

Chapter 14 Introduction to the Design Library 501

    Learn to select pre manufactured parts from the Design Library 507
    Learn to create a Hex Head bolt 509

Chapter 15 Introduction to the DMFXpress Analysis Wizard 515

    Learn to run an analysis on a part for manufacture 518
    Learn to interpret the results of the analysis 520
    Learn what steps to take to improve the design for manufacture process 522

Chapter 1 Getting Started

Objectives:
- Create a simple sketch using the Sketch commands
- Dimension a sketch using the Smart Dimension command
- Extrude a sketch using the Extruded Boss/Base command
- Create a hole using the Extrude Cut command
- Create a fillet using the Fillet command
- Create a counter bore using the Hole Wizard

Chapter 1 includes instruction on how to design the part shown below.

Chapter 1: Getting Started

1. Start SolidWorks by moving the cursor to the ![start] button in the lower left corner of the screen. Click the left mouse button once.

2. A pop up menu of the programs that are installed on the computer will appear. Scroll through the list of programs until you find "SolidWorks".

3. Move the cursor over the text "SolidWorks" and left click once as shown in Figure 1.

Figure 1

4. SolidWorks will open (load up and begin running).

5. Move the cursor to the upper left corner of the screen and left click on the "New Document" icon as shown in Figure 2.

Figure 2

Chapter 1: Getting Started

6. The New SolidWorks Document dialog box will appear. Left click on **Part** as shown in Figure 3.

Figure 3

7. Left click on **OK**.

3

Chapter 1: Getting Started

8. The screen should look similar to Figure 4. Left click on **Sketch**. If the Sketch tools are not visible, left click on the **Sketch** tab as shown.

Figure 4

Chapter 1: Getting Started

9. SolidWorks will display three different work planes. Move the cursor over the "Front Plane" until it turns red. Left click once as shown in Figure 5. SolidWorks will provide a perpendicular view of the front plane.

Figure 5

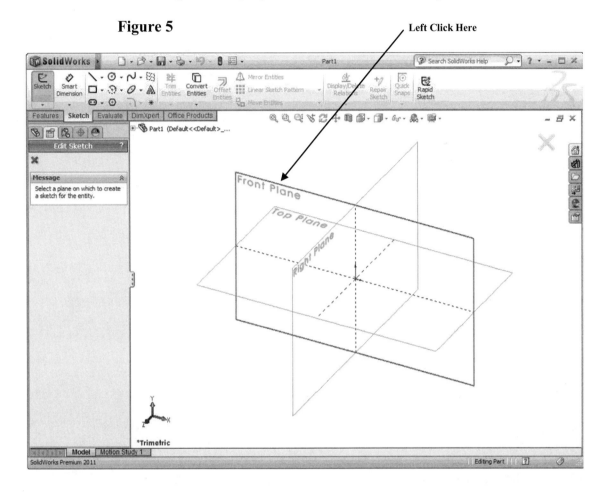

Chapter 1: Getting Started

10. Your screen should look similar to Figure 6.

Figure 6

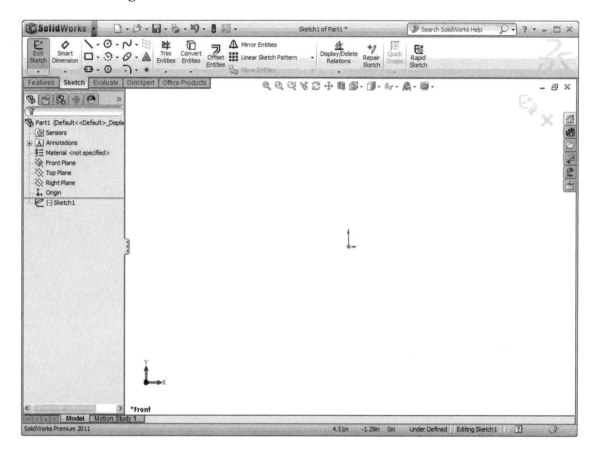

11. If the grid is displayed, skip to instruction number 12. If there is no grid visible, continue following the next steps.

Chapter 1: Getting Started

NOTE: Move the cursor to the upper left portion of the screen and left click on the arrow to the right of SolidWorks (the arrow will facing to the right). Left click on **Tools** as shown in Figure A.

Figure A

A drop down menu will appear. Left click on **Options** as shown in Figure B.

Figure B

Chapter 1: Getting Started

The Document Properties – Drafting Standard dialog box will appear. Left click on the **Document Properties** tab at the upper left portion of the dialog box as shown in Figure C.

Figure C

Move the cursor down to **Grid/Snap** and left click once. The Grid/Snap properties will appear. Place a check in the box (with the left mouse button) next to Display Grid as shown in Figure D. To change drawings units, (not shown) left click on **Units** and place a dot to the left of (IPS), inch pound, second.

Figure D

Chapter 1: Getting Started

Left click on **OK** as shown in Figure E.

Figure E

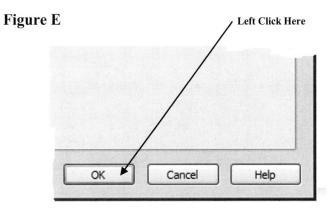

SolidWorks should now display a visible grid as shown in Figure F.

Figure F

Chapter 1: Getting Started

12. SolidWorks is now ready for use. Left click on **Line** as shown in Figure 7.

Figure 7

13. Move the cursor in the lower left portion of the screen and left click once. This will be the beginning end point of a line as shown in Figure 8.

Figure 8

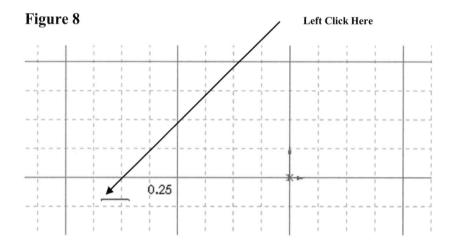

14. Move the cursor towards the lower right portion of the screen and left click once as shown in Figure 9.

Figure 9

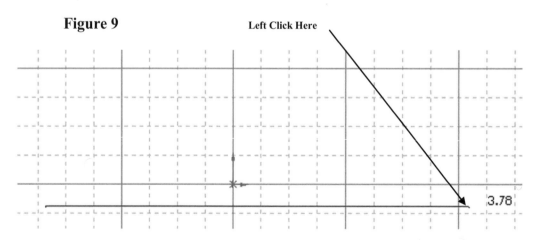

15. While the line is still attached to the cursor, move the cursor towards the top of the screen and left click once. Notice the length of the line is attached to the cursor at the right as shown in Figure 10.

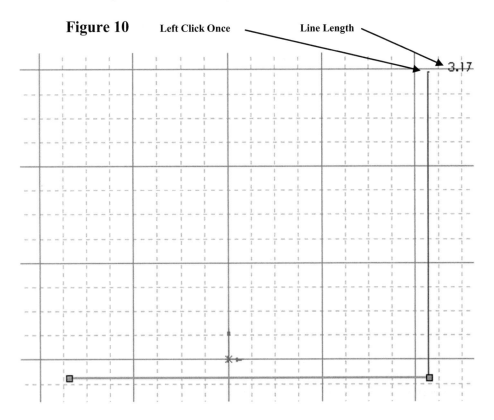

Chapter 1: Getting Started

16. With the line still attached to the cursor, move the cursor towards the left side of the screen. Notice the line of small dots connecting the first and last point together. Left click once when the small dots appear as shown in Figure 11.

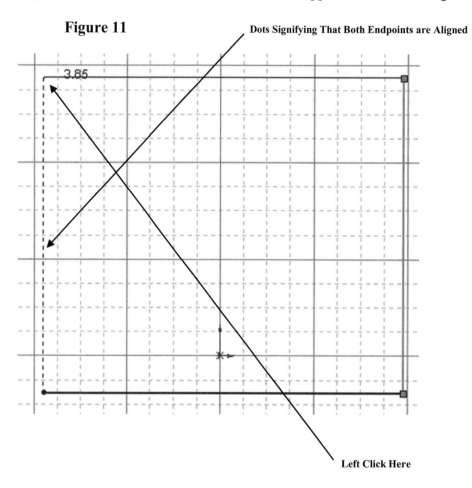

Figure 11

17. Your screen should look similar to Figure 12.

Figure 12

18. Move the cursor down towards the original starting point. Ensure that a red dot appears at the intersection of the two lines. This indicates that SolidWorks has "snapped" to the intersection of the lines. After the red dot appears, left click once. This will form a 90 degree box as shown in Figure 13.

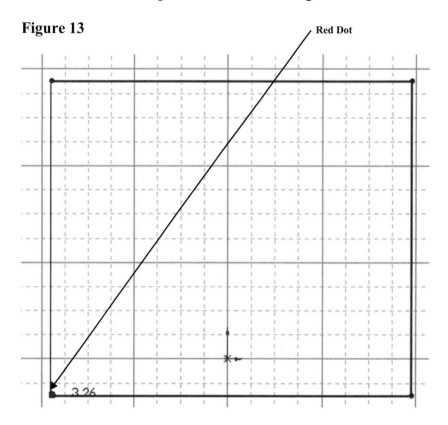

Figure 13

Chapter 1: Getting Started

19. Your screen should look similar to Figure 14. After you have closed the box, right click around the drawing. A pop up menu will appear. Left click on **Select** as shown in Figure 14. This verifies that no commands are active.

Figure 14

Chapter 1: Getting Started

20. Move the cursor to the upper left portion of the screen and left click on the arrow to the right of SolidWorks. Left click on **Tools**. A drop down menu will appear. Left click on **Options** located at the bottom of the menu as shown in Figure 15.

Figure 15

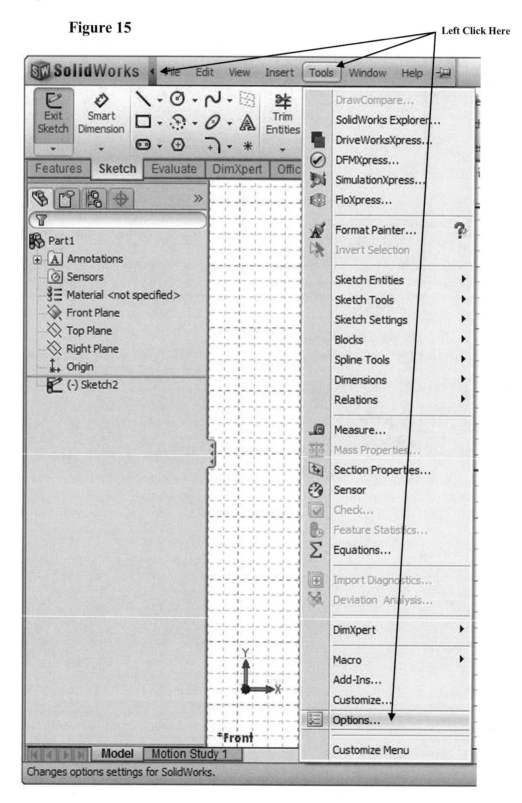

16

Chapter 1: Getting Started

21. The Document Properties – General dialog box will appear. Left click on the **Document Properties** tab. Left click on the drop down arrow to the right of "ANSI". A drop down menu will appear. Left click on **ANSI** as shown in Figure 16 (ANSI may already be selected).

Figure 16

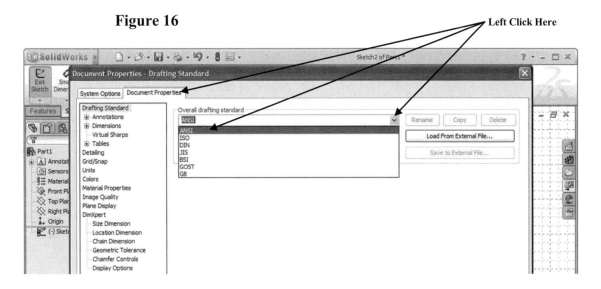

22. Left click on **OK** at the bottom of the screen.

Chapter 1: Getting Started

23. Move the cursor to the upper left portion of the screen and left click on **Smart Dimension** as shown in Figure 17.

Figure 17

24. Move the cursor over the bottom horizontal line until it turns red as shown in Figure 18. Select the line by left clicking anywhere on the line **or** on each of the end points. To use the end points of the line, move the cursor over one of the end points. A small red dot will appear. Left click once and move the cursor to the other end point. After the red dot appears, left click once. The dimension will be attached to the cursor.

Figure 18

25. Move the cursor down to where the dimension will be placed and left click once as shown in Figure 19.

Figure 19

18

26. The Modify dialog box will appear as shown in Figure 20.

Figure 20

Actual Dimension

27. To edit the dimension, enter **2.00** in the Modify dialog box (while the current dimension is highlighted) and left click on the green checkmark as shown in Figure 21.

Figure 21 Enter 2.00 Here Left Click Here

Chapter 1: Getting Started

28. The dimension of the line will become 2.00 inches as shown in Figure 22.

Figure 22

29. Move the cursor over the right side vertical line until it turns red as shown in Figure 23. Left click once.

Figure 23

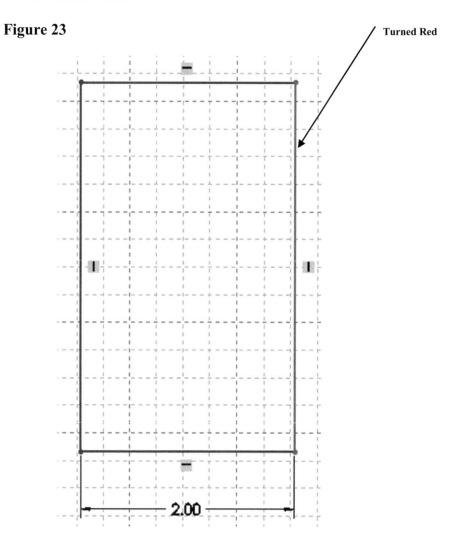

20

Chapter 1: Getting Started

30. Move the cursor to where the dimension will be placed and left click once as shown in Figure 24.

Figure 24

Chapter 1: Getting Started

31. The Modify dialog box will appear as shown in Figure 25.

Figure 25

32. To edit the dimension, enter **.25** in the Modify dialog box (while the current dimension is highlighted) and left click on the green checkmark as shown in Figure 26.

Figure 26

33. The screen should look similar to Figure 27.

Figure 27

Chapter 1: Getting Started

34. Right click anywhere around the drawing. A pop up menu will appear. Left click on **Select** as shown in Figure 28. This verifies that no commands are active.

Figure 28

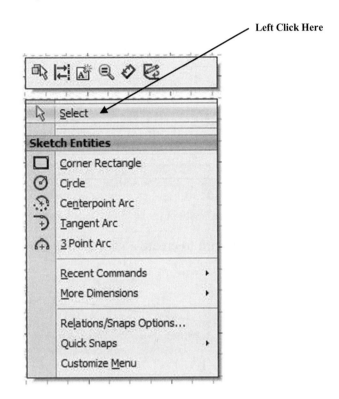

35. Move the cursor to the upper left corner of the screen and left click on **Line** as shown in Figure 29.

Figure 29

23

Chapter 1: Getting Started

36. Move the cursor to the upper left corner of the box as shown in Figure 30 and left click once.

Figure 30

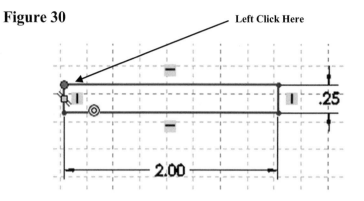

37. Move the cursor upward to create a vertical line and left click once as shown in Figure 31.

Figure 31

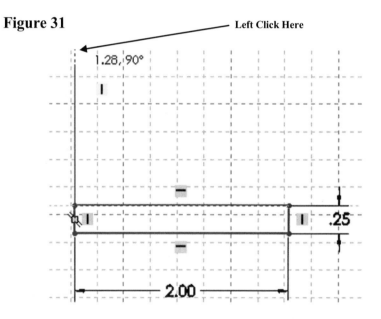

Chapter 1: Getting Started

38. Right click the mouse. A pop up menu will appear. Left click on **Select** as shown in Figure 32. This verifies that no commands are active.

Figure 32 — Left Click Here

39. You may have to zoom out to leave enough room to construct the following lines. To zoom out use the scroll wheel on the mouse or move the cursor to the upper middle portion of the screen and left click on the "Zoom to Fit" icon as shown in Figure 33.

Figure 33 — Left Click Here

40. Holding down the mouse wheel will allow you to rotate the sketch/part.

41. Move the cursor to the upper left corner of the screen and left click on **Line** as shown in Figure 34.

Figure 34 — Left Click Here

25

Chapter 1: Getting Started

42. Left click on the upper endpoint of the line. Move the cursor to the left and left click once as shown in Figure 35.

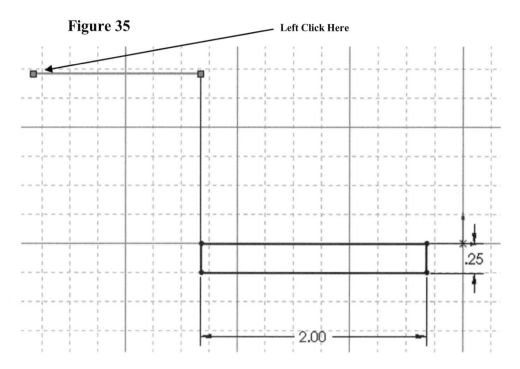

Figure 35

43. With the line still attached to the cursor, move the cursor up and left click once as shown in Figure 36.

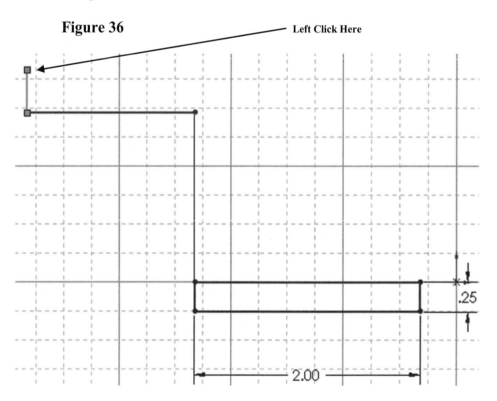

Figure 36

44. With the line still attached to the cursor, move the cursor to the right side of the screen and left click once as shown in Figure 37.

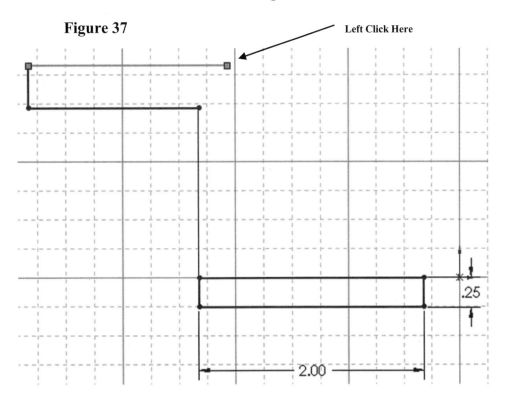

Figure 37

Chapter 1: Getting Started

45. With the line still attached to the cursor, move the cursor down towards the bottom of the screen and left click once as shown in Figure 38. Notice the small green square appearing at the intersection of the two lines. After the left mouse button is clicked the green square will change to a green dot. This indicates that the lines are all connected.

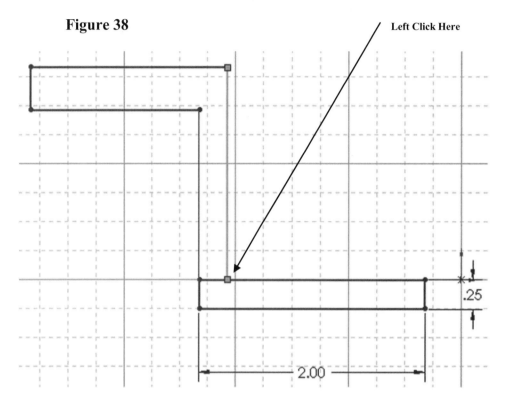

Figure 38

46. With the line still attached, right click the mouse. A pop up menu will appear. Left click on **Select** as shown in Figure 39. This verifies that no commands are active.

Figure 39

28

Chapter 1: Getting Started

47. Move the cursor to the upper middle portion of the screen and left click on the the arrow below Trim Entities as shown in Figure 40.

Figure 40

48. More icons will appear. Left click on **Trim Entities** as shown in Figure 40.

49. After selecting **Trim Entities** move the cursor to the middle left portion of the screen and left click on **Trim to closest**. Now move the cursor over the line that will be trimmed causing it to turn red and left click once as shown in Figure 41.

Figure 41

50. The line will disappear as shown in Figure 42. Left click on the green check mark under Trim or press the **Esc** key once. This verifies that no commands are active.

Figure 42

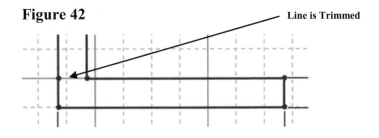

Chapter 1: Getting Started

51. Move the cursor over the line in the lower left corner of the drawing as shown in Figure 43. The line will turn red. This particular line must be deleted so that the line above can be extended the full length.

Figure 43

Line To Be Deleted

52. Left click once on the line that will be deleted. The line will turn red or green with small squares at each end as shown in Figure 44.

Figure 44

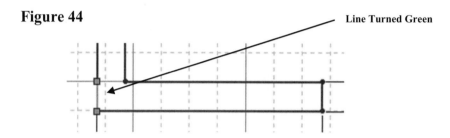

Line Turned Green

53. After the line is highlighted (turned red or green) right click once. A pop up menu will appear. Left click on **Delete** as shown in Figure 45.

Figure 45

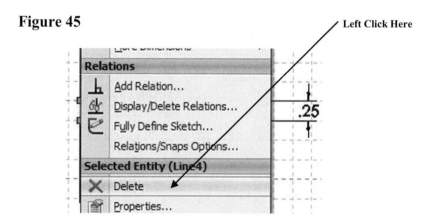

Left Click Here

54. The line will be deleted as shown in Figure 46.

Figure 46

Line is Deleted

Chapter 1: Getting Started

55. Move the cursor to the upper middle portion of the screen and left click on **Trim Entities** as shown in Figure 47.

Figure 47

56. There are several different methods of trimming or extending a line.

57. Move the cursor over the line to be extended. It will turn red as shown in Figure 48.

Figure 48

58. After the line has turned red, right click once. A pop up menu will appear. Left click on **Extend Entities** as shown in Figure 49.

Figure 49

31

Chapter 1: Getting Started

59. Move the cursor over the line to be extended. The line will extend downward forming one continuous line. Left click once as shown in Figure 50.

Figure 50

Chapter 1: Getting Started

60. Your screen should look similar to Figure 51.

Figure 51

61. Move the cursor to the upper left portion of the screen and left click on **Smart Dimension** as shown in Figure 52.

Figure 52

Chapter 1: Getting Started

62. After selecting **Smart Dimension** move the cursor over the left side vertical line. The line will turn red as shown in Figure 53. Left click on the line.

Figure 53

Line Turned Red

63. The dimension is attached to the cursor. Move the cursor to where the dimension will be placed and left click once. The Modify dialog box will appear as shown in Figure 54.

Figure 54

Actual Dimension

34

64. To edit the dimension, enter **1.75** in the Modify dialog box (while the current dimension is highlighted) and left click on the green checkmark as shown in Figure 55.

Figure 55

65. The dimension is now 1.75 as shown in Figure 56.

Figure 56

66. Move the cursor over the left side vertical line until it turns red as shown in Figure 57. Left click on the line.

Figure 57

67. The dimension is attached to the cursor. Move the cursor back and forth to verify it is attached. Move the cursor to where the dimension will be placed and left click once. The Modify dialog box will appear as shown in Figure 58.

Figure 58

Actual Dimension

68. To edit the dimension, enter **.25** in the Modify dialog box (while the current dimension is highlighted) and left click on the green checkmark as shown in Figure 59.

Figure 59

Enter .25 Here   Left Click Here

69. The dimension is now .25 as shown in Figure 60.

Figure 60

Chapter 1: Getting Started

70. Move the cursor over the top horizontal line until it turns red as shown in Figure 61. Left click on the line.

Figure 61

Turned Red

71. The dimension is attached to the cursor. Move the cursor up and down to verify it is attached. Move the cursor to where the dimension will be placed and left click once. The Modify dialog box will appear as shown in Figure 62.

Figure 62

Actual Dimension

72. To edit the dimension, enter **1.75** in the Modify dialog box (while the current dimension is highlighted) and left click on the green checkmark as shown in Figure 63.

Figure 63

Enter 1.75 Here   Left Click Here

37

Chapter 1: Getting Started

73. The dimension is now 1.75 as shown in Figure 64.

   Figure 64

74. Move the cursor over to the left side vertical line until it turns red and left click as shown in Figure 65.

   Figure 65

   Left Click Here

75. Move the cursor to the other vertical line until it turns red and left click as shown in Figure 66.

   Figure 66

   Left Click Here

Chapter 1: Getting Started

76. The dimension is attached to the cursor. Move the cursor back and forth to verify it is attached. Move the cursor to where the dimension will be placed and left click once. While the dimension is highlighted, left click the mouse once. The Modify dialog box will appear as shown in Figure 67.

Figure 67

77. To edit the dimension, enter **.25** in the Modify dialog box (while the current dimension is highlighted) and left click on the green checkmark as shown in Figure 68.

Figure 68

39

Chapter 1: Getting Started

78. Your screen should look similar to Figure 69.

Figure 69

79. Right click near the drawing. A pop up menu will appear. Left click on **Select** as shown in Figure 70. This will ensure that no commands are still active.

Figure 70

Left Click Here

Chapter 1: Getting Started

80. Move the cursor to the upper left portion of the screen and left click on **Exit Sketch** as shown in Figure 71.

Figure 71

81. Your screen should look similar to Figure 72.

Figure 72

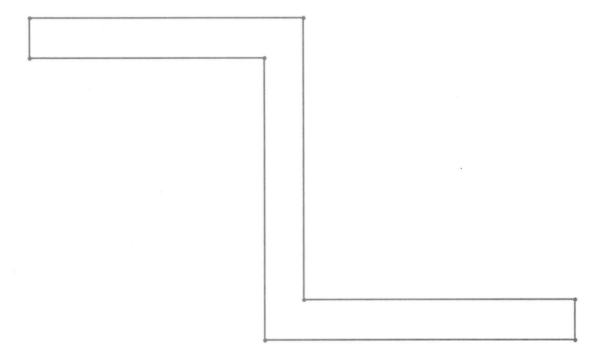

41

Chapter 1: Getting Started

82. To utilize the Features commands, a sketch must be present and have no opens (non-connected lines). If there are any opens in the sketch an error message will appear.

83. Move the cursor to the upper middle portion of the screen and left click on the drop down arrow next to the "View Orientation" icon as shown in Figure 73.

Figure 73

84. A drop down menu will appear. Left click on **Trimetric** as shown in Figure 74.

Figure 74

Chapter 1: Getting Started

85. The view will become trimetric. If needed use the scroll wheel to zoom out as shown in Figure 75.

Figure 75

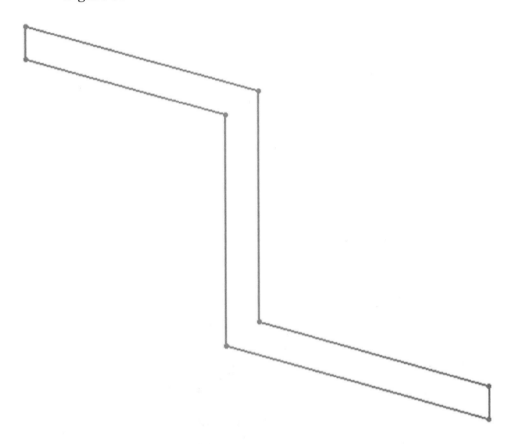

86. Move the cursor to the upper left portion of the screen and left click on **Extruded Boss/Base**. If the Extruded Boss/Base icon is not visible, left click on the **Features** tab as shown in Figure 76.

Figure 76

43

Chapter 1: Getting Started

87. The Extrude dialog box will appear. SolidWorks also provides a preview of the extrusion. If SolidWorks gave you an error message, there are opens (non-connected lines) somewhere on the sketch. Check each intersection for opens by using the Extend Entities and Trim Entities commands. Your screen should look similar to Figure 77. If the Extrude dialog box did not appear, left click on any line making up the sketch. To show shadows in the model left click on the drop down arrow to the right of the View Settings icon. A drop down menu will appear. Left click on **Shadows In Shaded Mode** as shown in Figure 77.

Figure 77

Chapter 1: Getting Started

88. Left click on the icon under Direction, located next to the text "Blind". The extrusion will change directions. Ensure that the extrusion is facing forward. While the text located next to "D1" is still highlighted, enter **1.000**. To preview the extruded part, left click on the "Glasses" icon located at the top of the Extrude dialog box as shown in Figure 78. The part will become purple.

Figure 78

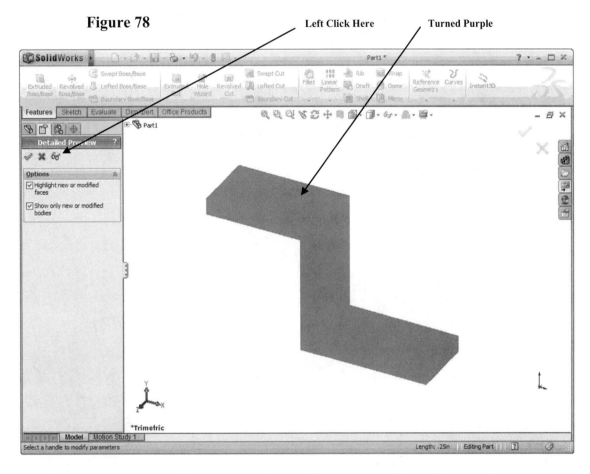

89. To end preview, left click on the "Glasses" icon again. This will return the part to the previous view.

90. Left click on the green checkmark as shown in Figure 79.

Figure 79

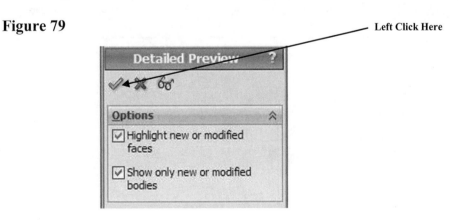

45

Chapter 1: Getting Started

91. Left clicking on the green checkmark will create a solid from the sketch. Your screen should look similar to Figure 80. Left click anywhere around the drawing.

Figure 80

92. Move the cursor to the upper middle portion of the screen and left click on **Fillet** as shown in Figure 81.

Figure 81

Chapter 1: Getting Started

93. The menu at the left will change. Enter **.50** for the radius and select **Full Preview** as shown in Figure 82.

Figure 82

Chapter 1: Getting Started

94. Move the cursor over the lower front edge causing it to turn red and left click once as shown in Figure 83.

Figure 83

Left Click Here

95. Notice the yellow lines illustrating a preview of the fillet as shown in Figure 84.

Figure 84

96. Move the cursor over the upper front edge causing it to turn red and left click once as shown in Figure 85.

Figure 85

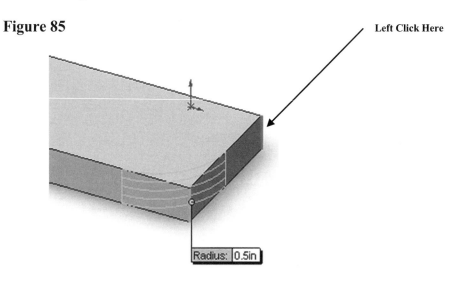

Left Click Here

97. Notice the yellow lines illustrating a preview of the fillet as shown in Figure 86.

Figure 86

98. Left click on the two upper remaining edges. You may have to use the rotate command to select the upper edge that is not visible.

99. **Note: The following is for users of SolidWorks 2009 and earlier. If you are using version 2010 then skip to step 100.** Move the cursor to the upper middle portion of the screen and <u>Right</u> click on the drop down arrow to the right of the View Orientation icon. A drop down menu will appear. Place a check in the box (with the left mouse button) next to the "Rotate" icon as shown in Figure 87. The Rotate icon will now appear with the Viewing tools located at the top of the screen in the drawing area. Once a check mark has been placed in the box, left click in the drawing area as shown in Figure 87. Pushing (and holding down) the scroll wheel while moving the mouse will also rotate the model.

Figure 87

Chapter 1: Getting Started

100. Left click on the left portion of the part (holding the left mouse button down). Drag the cursor to the right allowing access to the upper right corner of the part as shown in Figure 88.

Figure 88 — Left Click Here, Hold Left Mouse Button Down, and Drag to the Right

101. Move the cursor to the upper middle portion of the screen and left click on the "Rotate" icon again if the backside of the part is not visible as shown in Figure 89. Once the part is in the desired position hit the **ESC** key to escape out of the Rotate command.

Figure 89 — Left Click Here

Chapter 1: Getting Started

102. Move the cursor to the upper right corners of the part causing the edges to turn red. Left click once as shown in Figure 90.

Figure 90

103. Your screen should look similar to Figure 91.

Figure 91

104. Move the cursor to the upper middle portion of the screen and left click on the drop down arrow next to the "View Orientation" icon as shown in Figure 92.

Figure 92

Chapter 1: Getting Started

105. A drop down menu will appear. Left click on **Trimetric** as shown in Figure 93.

Figure 93

106. Your screen should look similar to Figure 94.

Figure 94

52

Chapter 1: Getting Started

107. Left click on the green checkmark as shown in Figure 95.

Figure 95

108. Your screen should look similar to Figure 96.

Figure 96

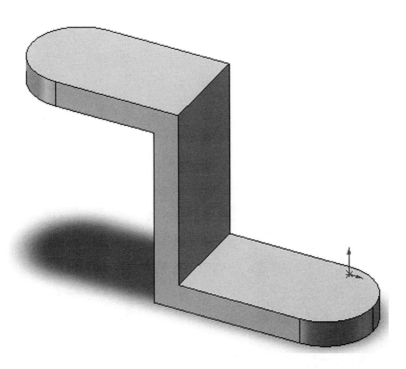

53

Chapter 1: Getting Started

109. The next task will include cutting a hole in each of the ends. To do this, a sketch will need to be constructed on each surface. To begin a new sketch on any surface, move the cursor to the surface that will include the new sketch. Notice the edges of the surface are red. After the edges have turned red, right click on the surface as shown in Figure 97.

Figure 97

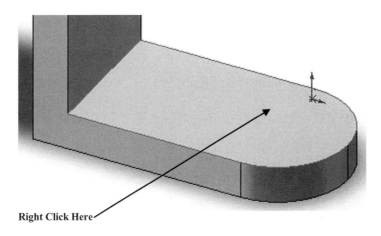

Right Click Here

110. The surface will change color. A pop up menu will appear. Left click on **Insert Sketch** as shown in Figure 98.

Figure 98

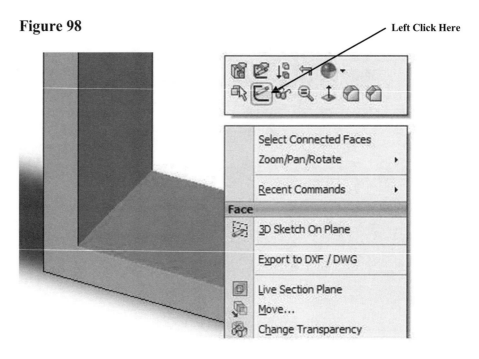

Left Click Here

111. SolidWorks will create a "sketch" on that particular surface. Notice the toolbar at the top of the screen has changed back to the sketch commands.

Chapter 1: Getting Started

112. Your screen should look similar to Figure 99.

   Figure 99

113. Move the cursor to the upper left portion of the screen and left click on **Circle** as shown in Figure 100.

   Figure 100   **Left Click Here**

114. Move the cursor to the edge of the circle. The center point of the circle will appear as shown in Figure 101.

   Figure 101   **Move Cursor Here**

   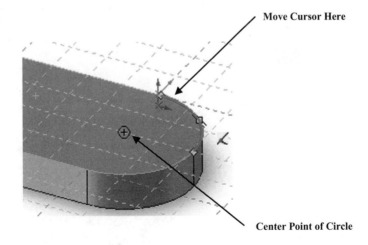

   Center Point of Circle

55

Chapter 1: Getting Started

115. After the center marker appears, left click on it once. This will be the center of a circle, which will later become a thru hole. Move the cursor out to the side to make the hole larger. Move the cursor out far enough to create a hole size similar to Figure 102.

Figure 102

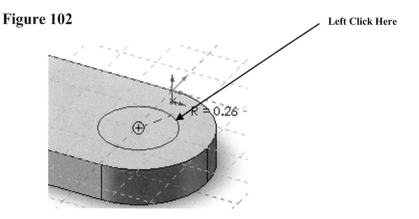

116. After the hole size looks similar to Figure 102, left click once. Left click on the green check mark at the middle left (not shown).

117. Move the cursor to the upper left portion of the screen and left click on **Smart Dimension** as shown in Figure 103.

Figure 103

118. Left click on the edge (not the center) of the circle as shown in Figure 104. The circle will turn red.

Figure 104

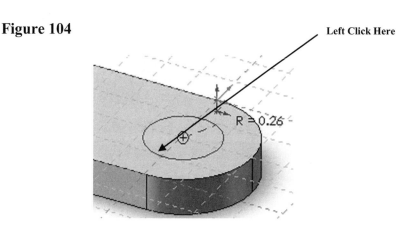

Chapter 1: Getting Started

119. The dimension will appear attached to the cursor. Move the cursor up and down to verify it is attached. Move the cursor to where the dimension will be placed and left click once as shown in Figure 105.

Figure 105

120. The Modify dialog box will appear as shown in Figure 106.

Figure 106

121. Enter **.500** in the Modify dialog box and left click on the green checkmark as shown in Figure 107.

Figure 107

57

Chapter 1: Getting Started

122. The diameter of the hole will become .500 inches as shown in Figure 108.

Figure 108

123. Right click near the drawing. A pop up menu will appear. Left click on **Select** as shown in Figure 109. This will ensure that no commands remain active.

Figure 109

124. Move the cursor to the upper left portion of the screen and left click on **Exit Sketch** as shown in Figure 110.

Figure 110

58

Chapter 1: Getting Started

125. SolidWorks is now out of the Sketch commands and into the Features Commands. Notice that the commands at the top of the screen are now different. Also notice that the Features tab is now active as shown in Figure 111.

Figure 111

126. Move the cursor to the upper left portion of the screen and left click on **Extruded Cut** as shown in Figure 112. This time SolidWorks will extrude "space" or "air" rather than material as was done to create the bracket. If the command does not become active you will need to left click on the circle that was created in the sketch area.

Figure 112

127. The menu at the left will change. Enter **.500** as shown in Figure 113.

Figure 113

59

Chapter 1: Getting Started

128.   Move the cursor to the upper left portion of the screen and left click on the "Glasses" icon as shown in Figure 114.

Figure 114

129.   A preview of the part will be displayed as shown in Figure 115.

Figure 115

130.   To return to the main menu, left click once on the "Glasses" icon.

131.   Left click on the green checkmark as shown in Figure 116.

Figure 116

132. There should be a thru hole in the part similar to Figure 117. Left click anywhere around the drawing

Figure 117

133. Another method of creating a hole is to use the Hole Wizard command.

134. To use the Hole Wizard command a "Point" must be constructed in the Sketch command. Move the cursor to the top portion of the part as shown in Figure 118. The outer edges of the part will turn red.

Figure 118

Move Cursor Here

Chapter 1: Getting Started

135. Right click on the surface. The surface will change color. A pop up menu will also appear. Left click on **Insert Sketch** as shown in Figure 119.

Figure 119

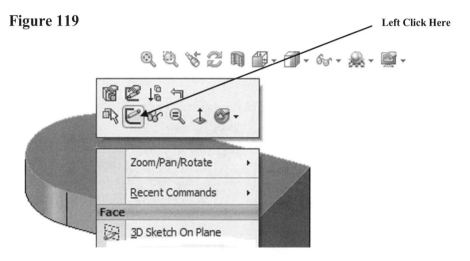

136. SolidWorks will return to the Sketch commands as shown in Figure 120.

Figure 120

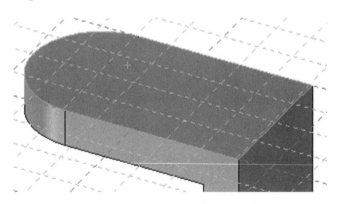

Chapter 1: Getting Started

137. Move the cursor to the upper right portion of the screen and left click on **Point** as shown in Figure 121.

Figure 121

138. Move the cursor to edge of the part. A portion of the edge will turn red. The center point of the fillet radius will also appear. Left click once on the center point as shown in Figure 122.

Figure 122

139. A small center marker will appear on the center of the fillet radius as shown in Figure 123.

Figure 123

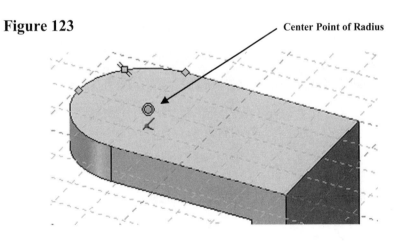

63

Chapter 1: Getting Started

140. Move the cursor to the upper left portion of the screen and left click on the green checkmark as shown in Figure 124.

Figure 124

141. Right click anywhere on the drawing. A pop up menu will appear. Left click on **Select** as shown in Figure 125.

Figure 125

142. Right click anywhere on the drawing. A pop up menu will appear. Left click on **Exit Sketch** as shown in Figure 126.

Figure 126

64

Chapter 1: Getting Started

143. SolidWorks is now out of the Sketch commands and into the Features commands. Notice that the commands at the top of the screen are now different. Your screen should look similar to Figure 127.

Figure 127

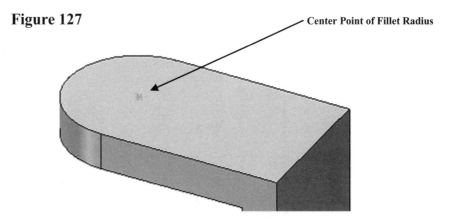

144. Move the cursor to the upper right portion of the screen and left click on **Hole Wizard** as shown in Figure 128.

Figure 128

Chapter 1: Getting Started

145. The Hole Specification dialog box will appear. Left click on the Counter Bore icon. Place a check in the "Show custom sizing" box. Below Custom Sizing enter **.250** for Hole Fit and Diameter, **.500** for C'Bore Diameter and **.125** for the C'Bore Depth as shown in Figure 129.

Figure 129

Chapter 1: Getting Started

146. Move the cursor to the top of the screen and left click on the **Positions** tab. Left click on **3D Sketch** as shown in Figure 130.

Figure 130

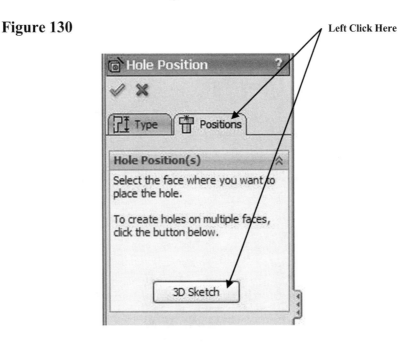

147. Move the cursor to the center point of the fillet radius and left click once as shown in Figure 131.

Figure 131

148. SolidWorks will provide a preview of the counterbore as shown in Figure 132.

Figure 132

67

Chapter 1: Getting Started

149. Left click on the green checkmark as shown in Figure 133. Left click anywhere around the drawing.

Figure 133

150. Your screen should look similar to Figure 134.

Figure 134

151. To ensure that the hole is correct move the cursor to the top portion of the screen and left click on the "Rotate" icon as shown in Figure 135.

Figure 135

68

152. Left click anywhere around the part and hold the left mouse button down. Drag the cursor upward. The part will rotate upward as shown in Figure 136.

Figure 136

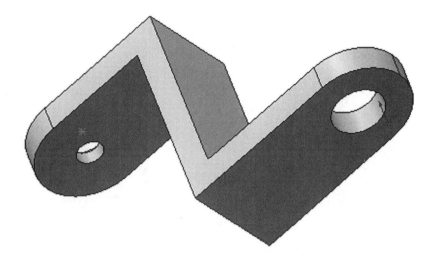

153. Holding the left mouse button down keeps the part attached to the cursor. To view the part in Trimetric, right click anywhere on the screen and left click on **View Orientation** as shown in Figure 137.

Figure 137

Left Click Here

Chapter 1: Getting Started

154. A drop down menu will appear. Double click on **Trimetric View** as shown in Figure 138.

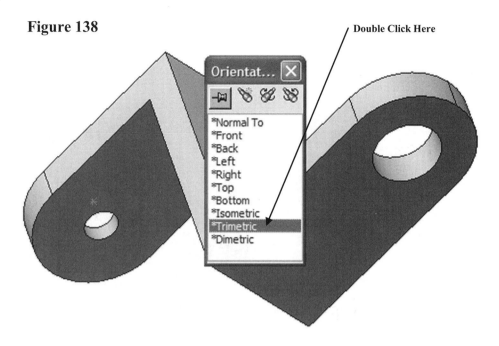

Figure 138

155. Other commands for viewing are located at the top of the screen as shown in Figure 139.

Figure 139

70

156. The Zoom to Area (Window) command works by using the cursor to draw a window around the area you want to zoom in on. After selecting the "Zoom Window" icon, hold the left mouse button down and drag a diagonal box around the desired area. Release the left mouse button when the proper amount of zoom is achieved.

157. The Zoom In/Out command works by scrolling the mouse wheel in either direction to achieve the desired view.

158. The Pan command works similar to the Zoom In/Out command. Start by selecting the "Pan" icon. If the Pan icon is not visible with the other Viewing tools it will need to be added by using the same steps outlined on page 49. Once the icon has been added to the Viewing tools, left click on the icon. Left click on the drawing and hold the left mouse button down while moving the cursor up and down or side to side. Release the mouse button after the desired view is achieved.

159. The Zoom to Fit command works by filling the screen so that the entire part is as large as possible while remaining visible in its entirety.

160. The same viewing options can be accessed by right clicking anywhere on the drawing. A pop will be displayed as shown in Figure 140.

Figure 140

Drawing Activities

Problem 1

Problem 2

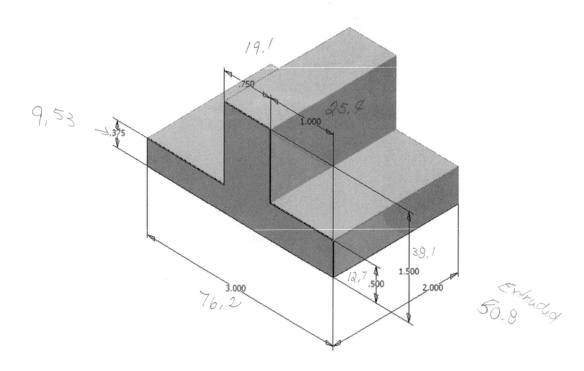

Problem 3

Extrude Center Section .25 Deep

Problem 4

Chapter 1: Getting Started

Problem 5

Problem 6

Problem 7

Problem 8

Chapter 2 Learning More Basics

Objectives:
- Create a simple sketch using the Sketch commands
- Dimension a sketch using the Smart Dimension command
- Revolve a sketch using the Revolved Boss/Base command
- Create a hole using the Extruded Cut command
- Create a series of holes using the Circular Pattern command

Chapter 2 includes instruction on how to design the part shown below.

Chapter 2: Learning More Basics

1. Start SolidWorks by referring to "Chapter 1 Getting Started".

2. After SolidWorks is running, begin a new sketch.

3. Move the cursor to the upper left corner of the screen and left click on **Line** as shown in Figure 1.

Figure 1

4. Move the cursor to the lower left portion of the screen and left click once. This will be the beginning end point of a line as shown in Figure 2.

Figure 2

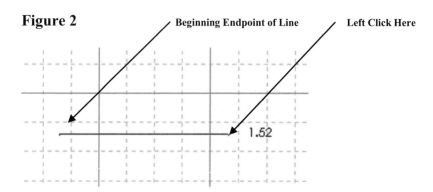

5. Move the cursor to the right and left click once as shown in Figure 2.

6. Move the cursor up and left click once as shown in Figure 3.

Figure 3

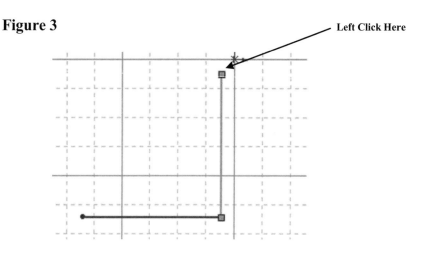

7. Move the cursor to the right and left click once as shown in Figure 4.

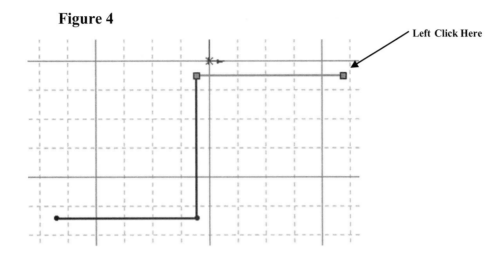

Figure 4

8. Move the cursor up and left click once as shown in Figure 5.

Figure 5

Chapter 2: Learning More Basics

9. Move the cursor to the left. Ensure that the dots between the first end point and the last end point appear as shown in Figure 6. Left click once.

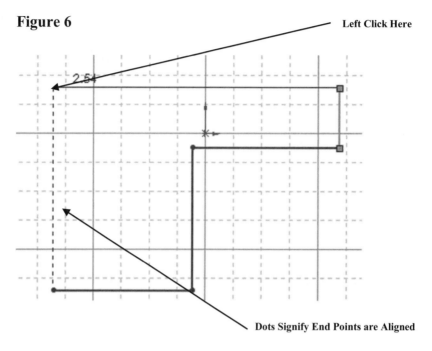

Figure 6

Left Click Here

Dots Signify End Points are Aligned

10. Move the cursor back to the original starting end point and left click once as shown in Figure 7. Press the **Esc** key once.

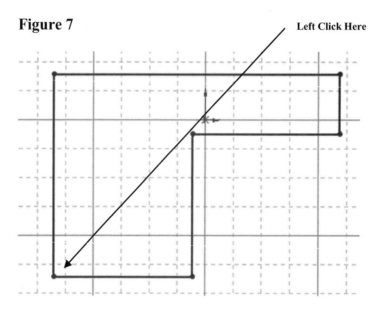

Figure 7

Left Click Here

Chapter 2: Learning More Basics

11. Move the cursor to the upper left portion of the screen and left click on **Smart Dimension** as shown in Figure 8.

Figure 8

12. After selecting **Smart Dimension** move the cursor over the bottom horizontal line until it turns red as shown in Figure 9. Select the line by left clicking anywhere on the line **or** on each of the end points. To use the end points of the line, move the cursor over one of the end points. A small red dot will appear. Left click once and move the cursor to the other end point. After the red dot appears, left click once. The dimension will be attached to the cursor.

Figure 9

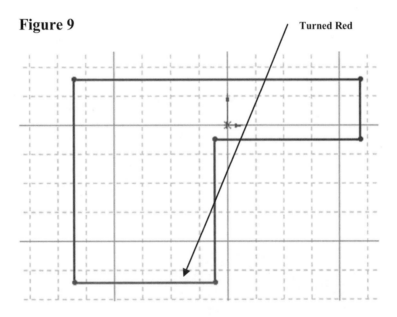

Chapter 2: Learning More Basics

13. Move the cursor to where the dimension will be placed and left click once as shown in Figure 10.

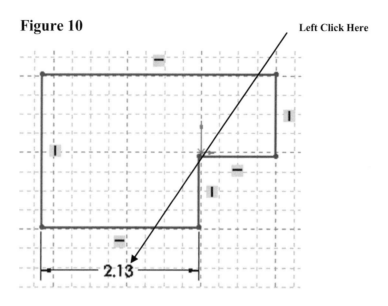

Figure 10

14. The Modify dialog box will appear as shown in Figure 11.

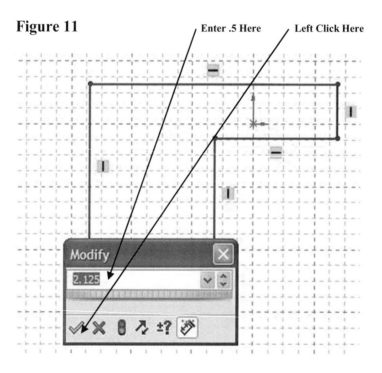

Figure 11

Chapter 2: Learning More Basics

15. To edit the dimension, enter **.50** in the Modify dialog box (while the current dimension is highlighted) and either press **Enter** on the keyboard or left click on the green checkmark as shown in Figure 11.

16. The dimension of the line will become .50 inches as shown in Figure 12.

Figure 12

17. To view the entire drawing move the cursor to the upper middle portion of the screen and left click once on the "Zoom To Fit" icon as shown in Figure 13.

Figure 13 ———————————————— Left Click Here

18. The drawing will "fill up" the entire screen. If the drawing is still too large, left click on the "Zoom In/Out" icon as shown in Figure 14. . If the Zoom In/Out icon is not visible with the other Viewing tools it will need to be added by using the same steps outlined on page 49. After selecting the icon, hold the left mouse button down and move the cursor up and down to achieve the desired view of the sketch.

Figure 14 ———————————————— Left Click Here

Chapter 2: Learning More Basics

19. Move the cursor to the middle left portion of the screen and left click on **Smart Dimension** as shown in Figure 15.

Figure 15

20. After selecting **Smart Dimension** move the cursor over the vertical line until it turns red as shown in Figure 16. Select the line by left clicking anywhere on the line **or** on each of the end points. To use the end points of the line, move the cursor over one of the end points. A small red dot will appear. Left click once and move the cursor to the other end point. After a red dot appears, left click once. The dimension will be attached to the cursor.

Figure 16

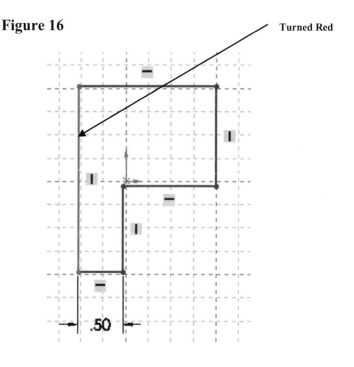

Chapter 2: Learning More Basics

21. Move the cursor to where the dimension will be placed and left click once as shown in Figure 17.

Figure 17

Left Click Here

22. The Modify dialog box will appear as shown in Figure 18.

Figure 18

Enter 2.0 Here   Left Click Here

23. To edit the dimension, enter **2.0** in the Modify dialog box (while the current dimension is highlighted) and either left click on the green checkmark (as shown in Figure 18) or press **Enter** on the keyboard.

24. The dimension of the line will become 2.0 inches as shown in Figure 19. Use the Zoom icons to zoom out if necessary.

Figure 19

25. Select the next line by left clicking anywhere on the line **or** on each of the end points. The dimension will be attached to the cursor.

Figure 20

Line Turned Red

Chapter 2: Learning More Basics

26. Move the cursor to where the dimension will be placed and left click once as shown in Figure 21.

Figure 21

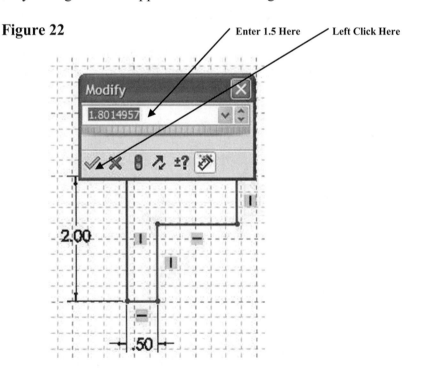

27. The Modify dialog box will appear as shown in Figure 22.

Figure 22

87

28. To edit the dimension, enter **1.5** in the Modify dialog box (while the current dimension is highlighted) and either left click on the green checkmark (as shown in Figure 22) or press **Enter** on the keyboard.

29. The dimension of the line will become **1.5** inches as shown in Figure 23. Use the Zoom icons to zoom out if necessary.

Figure 23

30. Select the next line by left clicking anywhere on the line **or** on each of the end points. The line will turn red as shown in Figure 24. The dimension will be attached to the cursor.

Figure 24

Chapter 2: Learning More Basics

31. Move the cursor to where the dimension will be placed and left click once as shown in Figure 25.

Figure 25

32. The Modify dialog box will appear as shown in Figure 26.

Figure 26

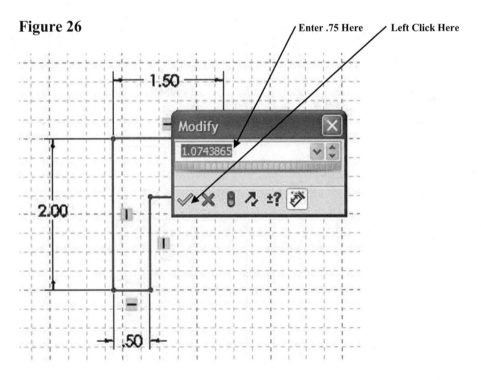

33. To edit the dimension, enter **.75** in the Modify dialog box (while the current dimension is highlighted) and either click on the green checkmark or press **Enter** on the keyboard.

89

Chapter 2: Learning More Basics

34. The dimension of the line will become **.75** inches as shown in Figure 27. Use the Zoom icons to zoom out if necessary.

Figure 27

35. Move the cursor to the upper left corner of the screen and left click on **Line** as shown in Figure 28.

Figure 28

Left Click Here

36. Draw a line parallel to the top horizontal line as shown in Figure 29.

Figure 29

Draw This Line

Chapter 2: Learning More Basics

37. Move the cursor to where the dimension will be placed and left click once. The Modify dialog box will appear. To edit the dimension, enter **.25** in the Modify dialog box (while the current dimension is highlighted) and either click on the green checkmark or press **Enter** on the keyboard. Your screen should look similar to Figure 30.

Figure 30

Dimension Line .25 From Line Below

38. Right click anywhere on the drawing. A pop up menu will appear. Left click on **Select** as shown in Figure 31. This verifies that no commands are active.

Figure 31

Left Click Here

39. After the sketch is complete it is time to revolve the sketch into a solid.

91

Chapter 2: Learning More Basics

40. Move the cursor to the upper left portion of the screen and left click on **Exit Sketch** as shown in Figure 32.

Figure 32

41. SolidWorks is now out of the Sketch commands and into the Features commands. Notice that the commands at the top of the screen are now different. To work in the Features commands a sketch must be present and have no opens (non-connected lines). If there are any opens in the sketch an error message will appear. Your screen should look similar to Figure 33.

Figure 33

92

Chapter 2: Learning More Basics

42. Move the cursor to the upper middle portion of the screen and left click on the drop down arrow next to the "View Orientation" icon. A drop down menu will appear as shown in Figure 34.

   Figure 34

43. Left click on **Trimetric** as shown in Figure 34.

44. The view will become trimetric as shown in Figure 35.

   Figure 35

Chapter 2: Learning More Basics

45. Move the cursor to the upper left portion of the screen and left click on **Revolved Boss/Base**. If the **Revolved Boss/Base** icon is not visible, left click on the **Features** tab as shown in Figure 36.

Figure 36

46. If the box to the right of the axis symbol contains text, move the cursor to the "axis" line that was drawn above the part and left click once. The line will turn red. If the box to the right of the axis symbol does not contain text, left click inside the box, then left click on the axis line as shown in Figure 37. It may take several attempts to complete the revolve.

Figure 37

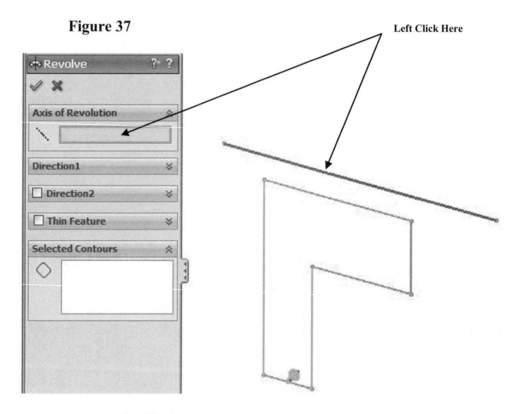

47. Move the cursor inside the profile causing it to turn red.

94

Chapter 2: Learning More Basics

48. Move the cursor under "Selected Contours" and left click once. Left click inside the profile as shown in Figure 38.

Figure 38

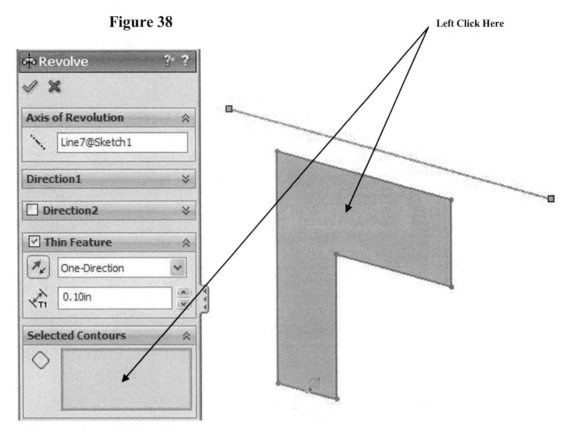

49. A preview of the revolve will appear as shown in Figure 39.

Figure 39

95

Chapter 2: Learning More Basics

50. Move the cursor to the middle left portion of the screen and left click on the checkmark next to the text "Thin Feature". **Remove the checkmark** (if needed) from the box as shown in Figure 40.

Figure 40

51. Move the cursor to the upper left portion of the screen and left click on the green checkmark as shown in Figure 41.

Figure 41

96

Chapter 2: Learning More Basics

52. Your screen should look similar to Figure 42. You may have to use the zoom out command to view the entire part. Left click anywhere around the drawing.

Figure 42

53. Move the cursor to the edge of the part causing the edges to turn red. After the edges become red, right click on the surface as shown in Figure 43.

Figure 43   Right Click Here

97

Chapter 2: Learning More Basics

54. The surface will turn green and a pop up menu will appear. Left click on **Insert Sketch** as shown in Figure 44.

Figure 44

55. SolidWorks will begin a new sketch on the selected surface. Your screen should look similar to Figure 45.

Figure 45

98

Chapter 2: Learning More Basics

56. Move the cursor to the upper right portion of the screen and left click on the drop down arrow next to the "View Orientation" icon. A drop down menu will appear. Left click on **Right** as shown in Figure 46.

Figure 46

57. SolidWorks will rotate the part to provide a perpendicular view of the selected surface as shown in Figure 47.

Figure 47

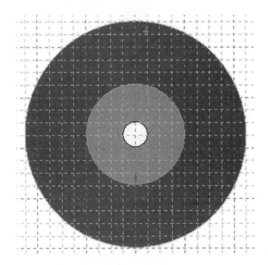

58. Move the cursor to the upper left corner of the screen and left click on **Line** as shown in Figure 48.

Figure 48

Chapter 2: Learning More Basics

59. Move the cursor over the small circle and wait a few seconds. A center marker will appear. Left click on the center marker as shown in Figure 49.

Figure 49

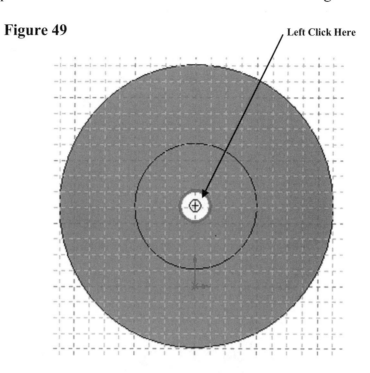

60. Move the cursor straight up and left click as shown in Figure 50.

Figure 50

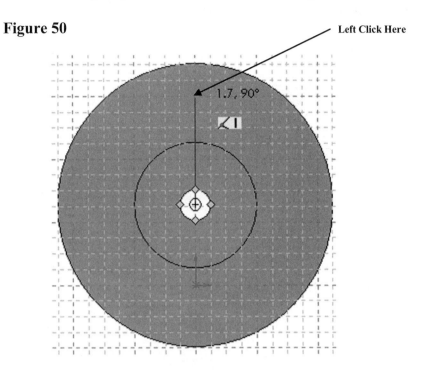

61. Now right click. A pop up menu will appear. Left click on **Select** as shown in Figure 51.

Figure 51

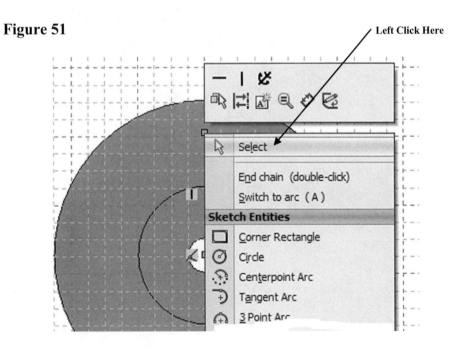

Chapter 2: Learning More Basics

62. Move the cursor to the upper left portion of the screen and left click on **Smart Dimension** as shown in Figure 52.

Figure 52

Left Click Here

63. Move the cursor over the line that was just drawn. The line will turn red as shown in Figure 53. Select the line by left clicking anywhere on the line **or** on each of the end points. The dimension will be attached to the cursor.

Figure 53

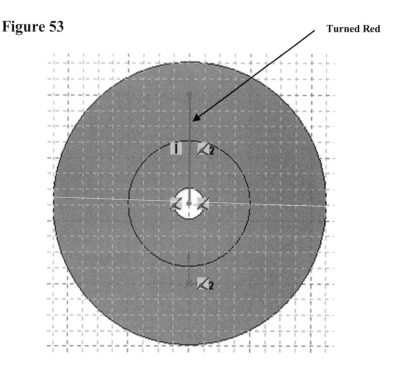

Turned Red

102

64. Move the cursor to where the dimension will be placed and left click once as shown in Figure 54.

Figure 54

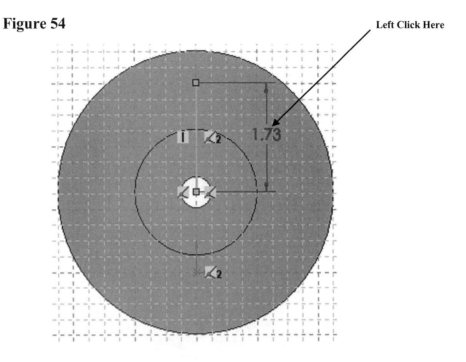

65. The Modify dialog box will appear as shown in Figure 55.

Figure 55

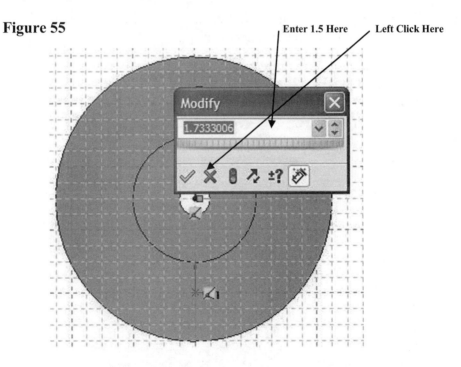

66. To edit the dimension, enter **1.5** in the Modify dialog box (while the current dimension is highlighted) as shown in Figure 55. Left click on the green checkmark or press **Enter** on the keyboard.

Chapter 2: Learning More Basics

67. Right click anywhere on the screen. A pop up menu will appear. Left click on **Select** as shown in Figure 56.

Figure 56

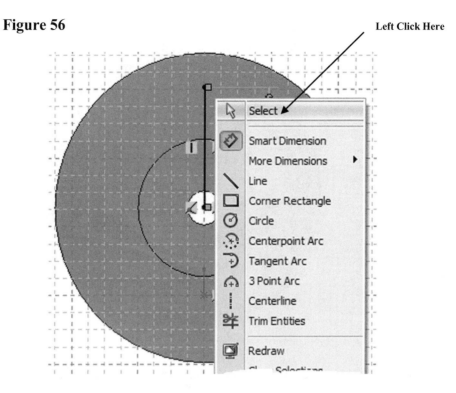

68. The dimension of the line will become 1.5 inches as shown in Figure 55. Use the Zoom icons to zoom out if necessary.

Figure 57

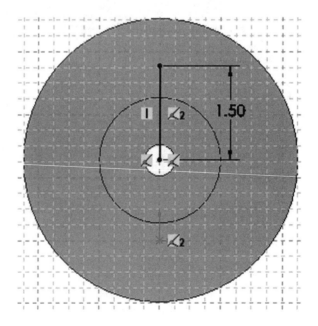

104

Chapter 2: Learning More Basics

69. Move the cursor to the upper left portion of the screen and left click on **Circle** as shown in Figure 58.

Figure 58

70. Left click on the endpoint of the line as shown in Figure 59.

Figure 59

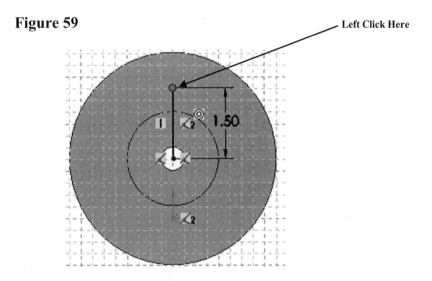

71. Move the cursor out to create a circle as shown in Figure 60.

Figure 60

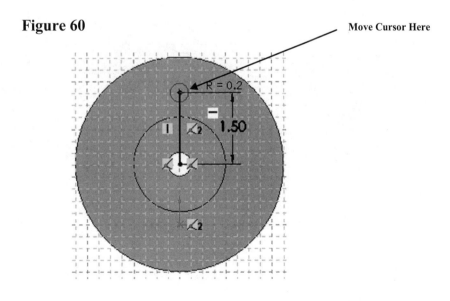

105

Chapter 2: Learning More Basics

72. Left click as shown in Figure 61.

Figure 61

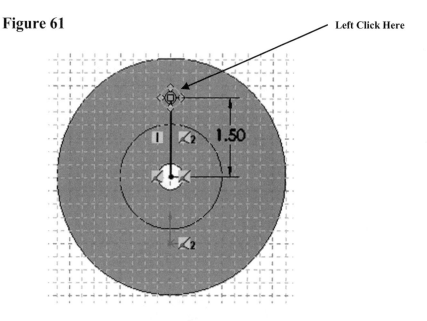

73. Move the cursor to the left portion of the screen and left click on the green checkmark as shown in Figure 62.

Figure 62

74. Move the cursor to the upper left portion of the screen and left click on **Smart Dimension** as shown in Figure 63.

Figure 63

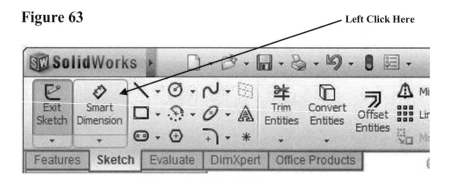

106

75. Move the cursor to the circle that was just drawn. The circle will turn red. Select the circle by left clicking anywhere on the circle (not the center) as shown in Figure 64. The dimension will be attached to the cursor.

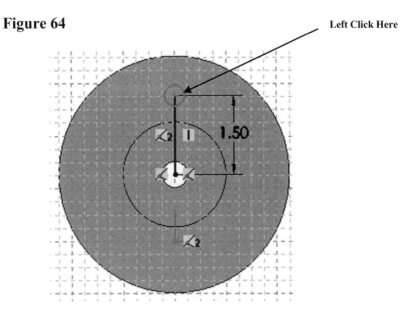

Figure 64

76. Move the cursor to where the dimension will be placed and left click once as shown in Figure 65.

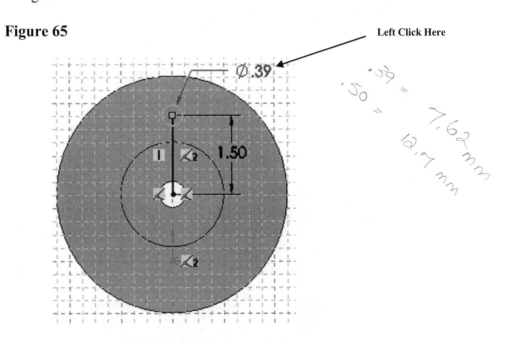

Figure 65

Chapter 2: Learning More Basics

77. The Modify dialog box will appear as shown in Figure 66.

   Figure 66

   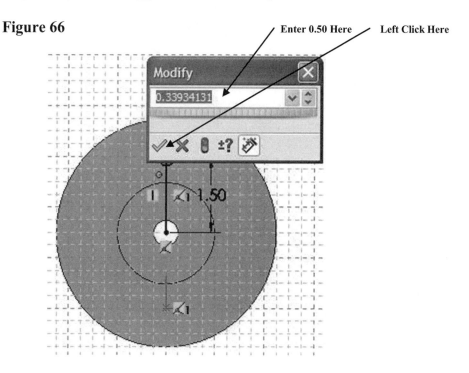

78. To edit the dimension, enter **0.50** in the Modify dialog box (while the current dimension is highlighted) and either press **Enter** on the keyboard or left click on the green checkmark as shown in Figure 66.

79. The dimension of the line will become **0.50** inches as shown in Figure 67. Use the Zoom icons to zoom out if necessary.

   Figure 67

   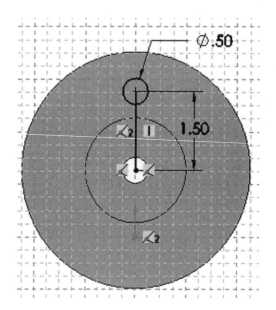

108

Chapter 2: Learning More Basics

80. Right click anywhere on the screen. A pop up menu will appear. Left click on **Select** as shown in Figure 68.

Figure 68

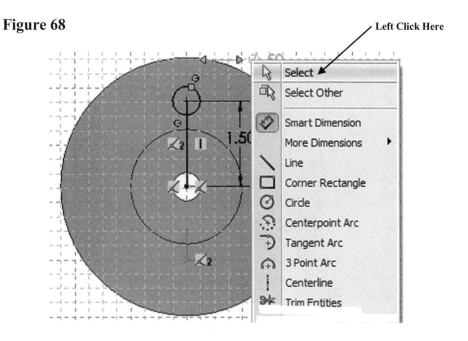

81. Move the cursor to the line that was used to locate the center of the circle. The line will turn red as shown in Figure 69.

Figure 69

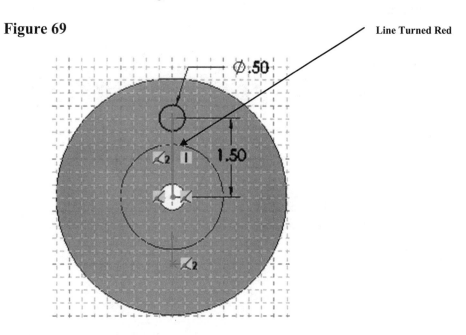

109

Chapter 2: Learning More Basics

82. Right click on the line after it turns red. A pop up menu will appear as shown in Figure 70.

Figure 70

83. Left click on **Delete** as shown in Figure 70.

84. The Sketcher Confirm Delete dialog box will appear. Left click on **Yes** as shown in Figure 71.

Figure 71

85. After you have verified that no commands are active, move the cursor to the upper left portion of the screen and left click on **Exit Sketch** as shown in Figure 72.

   Figure 72

86. SolidWorks is now out of the Sketch commands and into the Features commands. Notice that the commands at the top of the screen are now different. To work in the Features commands a sketch must be present and have no opens (non-connected lines). If there are any opens in the sketch an error message will appear. Your screen should look similar to Figure 73.

   Figure 73

Chapter 2: Learning More Basics

87. Move the cursor to the upper right portion of the screen and left click on the drop down arrow next to the "Views Orientation" icon. A drop down menu will appear. Left click on **Trimetric** as shown in Figure 74.

Figure 74

88. The view will become trimetric as shown in Figure 75.

Figure 75

112

Chapter 2: Learning More Basics

89. Move the cursor to the upper middle portion of the screen and left click on **Extruded Cut**. If the **Extruded Cut** icon is not visible, left click on the **Features** tab as shown in Figure 76.

Figure 76

90. If SolidWorks did not provide a preview of the Extruded Cut, move the cursor to the edge of the circle. The edge will turn red as shown in Figure 77.

Figure 77

91. After the hole turns red, left click once.

92. A preview of the hole is displayed as shown in Figure 78.

Figure 78

Arrow is Short and Depth of Hole is Shallow

93. Move the cursor to the middle left portion of the screen. Highlight the text located next to D1 (as shown in Figure 79) and enter **.75** and press the enter key on the keyboard.

Figure 79

Enter .75 Here

Chapter 2: Learning More Basics

94. Notice that the depth of the hole has changed in the preview. The arrow illustrating the hole depth is now longer as shown in Figure 80.

Figure 80

Longer Arrow Means Hole Depth is Deeper

95. Left click on the green checkmark as shown in Figure 81.

Figure 81

Left Click Here

115

96. Your screen should look similar to Figure 82.

Figure 82 — Thru Hole

97. Move the cursor to the far right portion of the screen and left click on the arrow located below Reference Geometry as shown in Figure 83.

Figure 83 — Left Click Here

98. A drop down menu will appear. Left click on **Axis** as shown in Figure 84.

Figure 84 — Left Click Here

Chapter 2: Learning More Basics

99. Move the cursor over the center of the center hole causing it to turn red as shown in Figure 85.

Figure 85

Turned Red

100. After the inside of the hole turns red, left click once. A center axis will be placed inside the hole as shown in Figure 86.

Figure 86

Center Axis

Chapter 2: Learning More Basics

101. Move the cursor to the middle left portion of the screen and left click on the green checkmark as shown in Figure 87.

Figure 87

102. Move the cursor to the upper middle portion of the screen and left click on the drop down arrow located below Linear Pattern. A drop down menu will appear. Left click on **Circular Pattern** as shown in Figure 88.

Figure 88

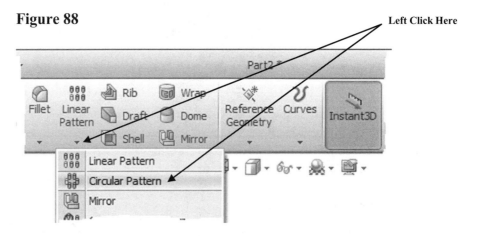

103. Move the cursor to the inside edge of the hole. The outside edge of the entire hole will turn red. Left click once. Left click on the axis as shown in Figure 89.

Figure 89

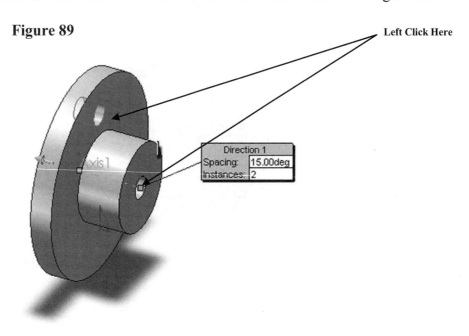

104. Move the cursor to the left portion of the screen. Place a checkmark in the box next to the text "Equal Spacing". The degree amount automatically changes to 360 degrees. Enter **6** for the number of holes directly below 360 and press **Enter** on the keyboard as shown in Figure 90.

Figure 90

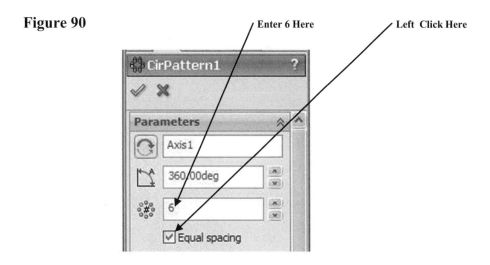

105. A preview of the circular pattern will be displayed as shown in Figure 91.

Figure 91

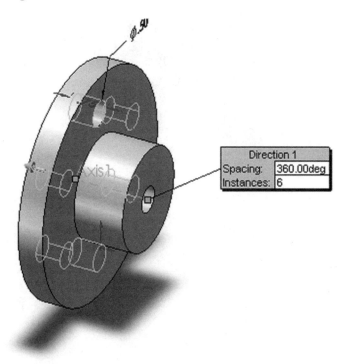

Chapter 2: Learning More Basics

106. Move the cursor to the upper left portion of the screen and left click on the green checkmark as shown in Figure 92.

Figure 92

107. Your screen should look similar to Figure 93.

Figure 93

Chapter 2: Learning More Basics

108. To ensure that the holes are correct, move the cursor to the upper middle portion of the screen and left click on the "Rotate" icon. If the Rotate icon is not visible (2009 and earlier users) move the cursor to the upper middle portion of the screen and Right click on the drop down arrow to the right of the View Orientation icon. A drop down menu will appear. Place a check in the box (with the left mouse button) next to the "Rotate" icon as shown in Figure 94. The Rotate icon will now appear with the Viewing tools located at the top of the screen in the drawing area. The Rotate command can also be made active by pressing the mouse wheel down and moving the cursor to achieve the desired view.

Figure 94

121

Chapter 2: Learning More Basics

109. The part will rotate as shown in Figure 95.

Figure 95

110. Right click around the part. A pop up menu will appear. Left click on **Select**. Other viewing options are shown in the pop up menu in Figure 96.

Figure 96

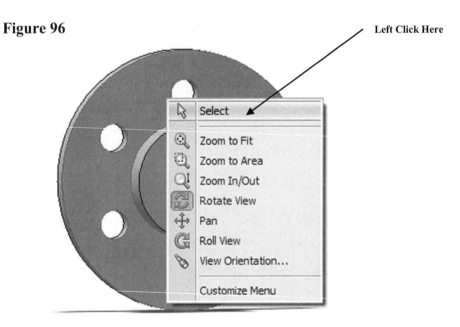

Left Click Here

Drawing Activities

Problem 1

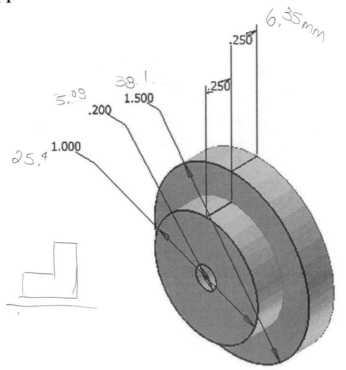

3.500 " = 88.9 mm
3.000 " = 76.20 mm
2.500 " = 63.5 mm
2.000 " = 50.8 mm
1.500 " = 38.1 mm
1.000 inch = 25.4 mm
0.200 inch = 5.09 mm
0.250 inch = 6.35 mm

.500 inch = 12.7 mm

Problem 2

Chapter 2: Learning More Basics

Problem 3

Problem 4

Problem 5

Problem 6

Chapter 2: Learning More Basics

Problem 7

Problem 8

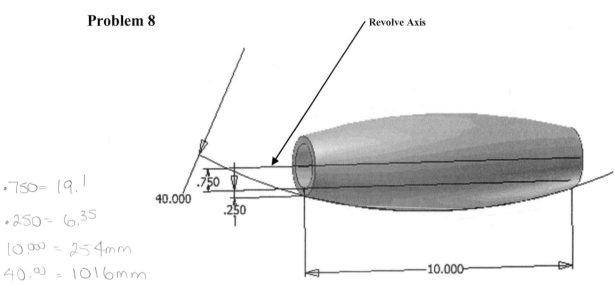

.750 = 19.1
.250 = 6.35
10.000 = 254mm
40.00 = 1016mm

Chapter 3 Learning To Create a Detail Drawing

Objectives:
- Create a simple sketch using the Sketch commands
- Extrude a sketch into a solid using the Features commands
- Create an Orthographic view using the Drawing commands
- Create a Solid Model view using the Model View command

Chapter 3 includes instruction on how to design the parts shown below.

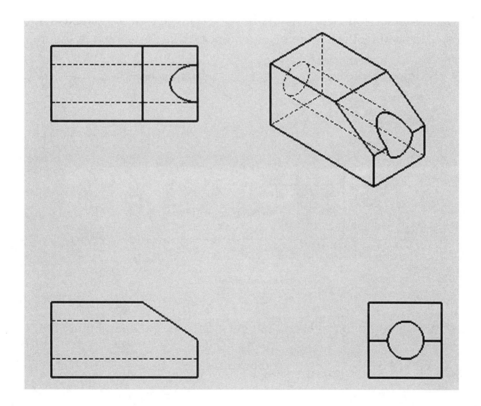

Chapter 3: Learning To Detail Part Drawings

1. Start SolidWorks by referring to "Chapter 1 Getting Started".

2. After SolidWorks is running, begin a new sketch.

3. Move the cursor to the upper left corner of the screen and left click on **Line** as shown in Figure 1.

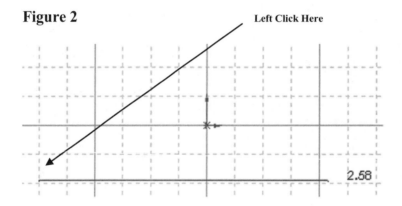

Figure 1

4. Move the cursor somewhere in the lower left portion of the screen and left click once. This will be the beginning end point of a line as shown in Figure 2.

Figure 2

5. Move the cursor to the right and left click once as shown in Figure 3.

Figure 3

Chapter 3: Learning To Detail Part Drawings

6. Move the cursor upward and left click once as shown in Figure 4.

Figure 4

7. Move the cursor to the left and wait for dots to appear. Left click once as shown in Figure 5.

Figure 5

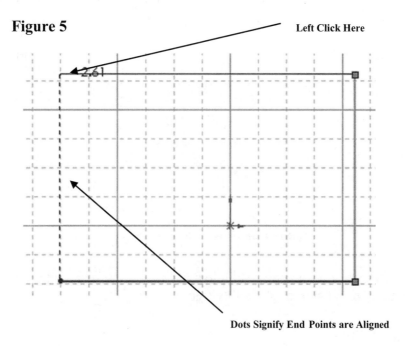

Chapter 3: Learning To Detail Part Drawings

8. Move the cursor back to the original starting end point. A red dot will appear. Left click once. Your screen should look similar to Figure 6.

9. Right click anywhere on the screen. A pop up menu will appear. Left click on **Select** as shown in Figure 7.

10. Move the cursor to the upper middle portion of the screen and left click on **Smart Dimension** as shown in Figure 8.

Chapter 3: Learning To Detail Part Drawings

11. After selecting **Smart Dimension** move the cursor over the bottom horizontal line until it turns red as shown in Figure 9. Select the line by left clicking anywhere on the line **or** on each of the end points. The dimension will be attached to the cursor.

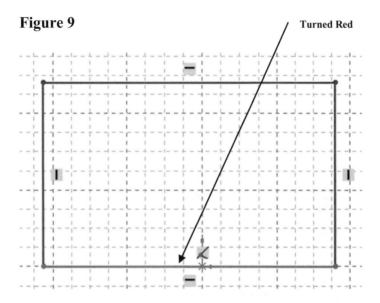

Figure 9 — Turned Red

12. Move the cursor to where the dimension will be placed and left click once as shown in Figure 10.

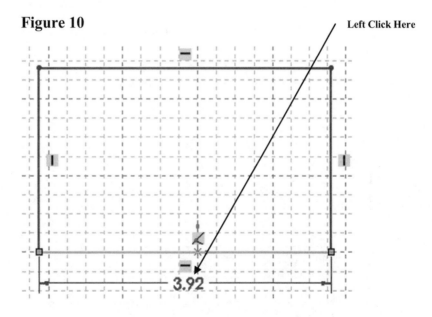

Figure 10 — Left Click Here

131

Chapter 3: Learning To Detail Part Drawings

13. The Modify dialog box will appear as shown in Figure 11.

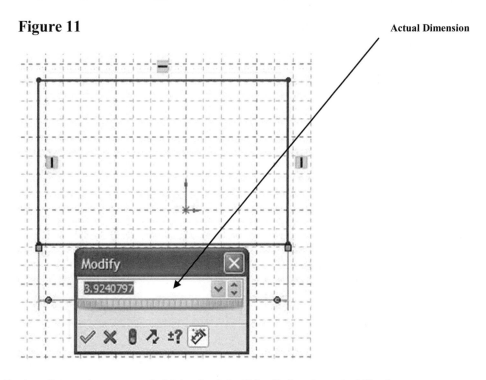

Figure 11   Actual Dimension

14. To edit the dimension, enter **2.00** in the Modify dialog box (while the current dimension is highlighted) and either press **Enter** on the keyboard or left click on the green checkmark as shown in Figure 12.

Figure 12   Enter 2.00 Here   Left Click Here

15. The dimension of the line will become 2.00 inches as shown in Figure 13.

Figure 13

16. Right click anywhere on the screen. A pop up menu will appear. Left click on **Select** as shown in Figure 14.

Figure 14   Left Click Here

17. To view the entire drawing, move the cursor to the upper right portion of the screen and left click once on the "Zoom to Fit" icon as shown in Figure 15.

Figure 15   Left Click Here

Chapter 3: Learning To Detail Part Drawings

18. The drawing will "fill up" the entire screen. If the drawing is still too large, left click on the "Zoom In/Out" icon as shown in Figure 16. Hold the left mouse button down and move the cursor up and down to achieve the desired view of the sketch.

Figure 16

19. Move the cursor to the upper left portion of the screen and left click on **Smart Dimension** as shown in Figure 17.

Figure 17

20. Move the cursor over the right side line until it turns red as shown in Figure 18. Select the line by left clicking anywhere on the line **or** on each of the end points. The dimension will be attached to the cursor.

Figure 18

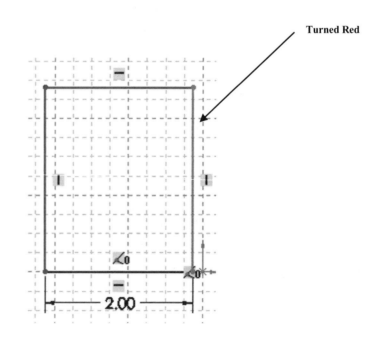

134

21. Move the cursor to where the dimension will be placed and left click once as shown in Figure 19.

Figure 19   **Left Click Here**

22. The Modify dialog box will appear as shown in Figure 20.

Figure 20   **Actual Dimension**

Chapter 3: Learning To Detail Part Drawings

23. To edit the dimension, enter **1.00** in the Modify dialog box (while the current dimension is highlighted) and either press **Enter** on the keyboard or left click on the green checkmark as shown in Figure 21.

Figure 21

24. The dimension of the line will become 1.00 inches as shown in Figure 22. Use the Zoom icons to zoom out if necessary.

Figure 22

25. Right click anywhere on the screen. A pop up menu will appear. Left click on **Select** as shown in Figure 23.

Figure 23

Chapter 3: Learning To Detail Part Drawings

26. Move the cursor to the upper left portion of the screen and left click on **Smart Dimension** as shown in Figure 24.

Figure 24

27. After selecting **Smart Dimension** move the cursor over the top horizontal line until it turns red as shown in Figure 25. Select the line by left clicking anywhere on the line **or** on each of the end points. This will cause the dimension to be attached to the cursor.

Figure 25

28. Move the cursor to where the dimension will be placed and left click once as shown in Figure 26.

Figure 26

137

Chapter 3: Learning To Detail Part Drawings

29. Notice that the dimension is exactly 2.000. Move the cursor to where the dimension will be placed and left click once. The Make Dimension Driven? dialog box will appear as shown in Figure 27.

Figure 27

30. Right click anywhere on the screen. A pop up menu will appear. Left click on **Select** as shown in Figure 28.

Figure 28

31. This dimension will over constrain the sketch because the sketch has been constrained with 90 degree angles when it was constructed. Left click on **OK**. The dimension will be driven meaning it cannot be used to edit or change the length of the line.

138

32. The driven dimension appears grayed out as shown in Figure 29.

Figure 29

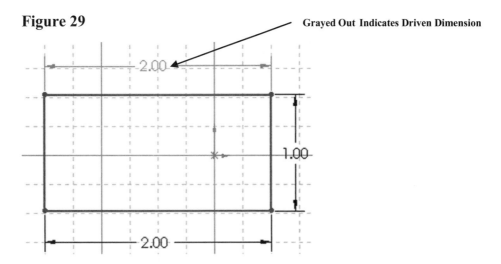

Grayed Out Indicates Driven Dimension

33. Dimensioning the far left line would also result in a driven dimension. Because of this, the dimensioning portion is complete.

34. After the sketch is complete it is time to extrude the sketch into a solid.

35. After you have verified that no commands are active, move the cursor to the upper left portion of the screen and left click on **Exit Sketch** as shown in Figure 30.

Figure 30

Left Click Here

Chapter 3: Learning To Detail Part Drawings

36. SolidWorks is now out of the Sketch commands and into the Features commands. Notice that the commands at the top of the screen are now different. To work in the Features commands a sketch must be present and have no opens (non-connected lines). If there are any opens in the sketch an error message will appear. Your screen should look similar to Figure 31.

Figure 31

37. Move the cursor to the upper middle portion of the screen and left click on the drop down arrow next to the "Views Orientation " icon. A drop down menu will appear. Left click on **Trimetric** as shown in Figure 32.

Figure 32　　　　　　　　　　　　　　　　　　　　　**Left Click Here**

140

38. The view will become trimetric as shown in Figure 33.

Figure 33

39. Move the cursor to the upper left portion of the screen and left click on **Extruded Boss/Base** as shown in Figure 34. If SolidWorks gave you an error message, there are opens (non-connected lines) somewhere on the sketch. Check each intersection for opens by using the **Extend** and **Trim** commands.

Figure 34

Left Click Here

Chapter 3: Learning To Detail Part Drawings

40. A preview of the extrusion will be displayed as shown in Figure 35.

Figure 35

41. If a preview does not appear, move the cursor over any of the lines. The line will turn red as shown in Figure 36. Left click once.

Figure 36 Turned Red

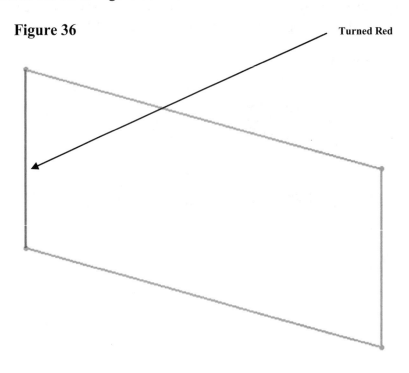

142

42. Enter **1.00** for D1 as shown in Figure 37.

Figure 37

Enter 1.00 Here

43. Move the cursor to the upper left portion of the screen and left click on the green checkmark as shown in Figure 38. Left click once anywhere around the drawing.

Figure 38

Left Click Here

Chapter 3: Learning To Detail Part Drawings

44. Your screen should look similar to Figure 39. You may have to use the Zoom Out command to view the entire part.

Figure 39

45. Move the cursor to the upper middle portion of the screen and left click on the drop down arrow located under Fillet. A drop down menu will appear. Left click on **Chamfer** as shown in Figure 40.

Figure 40

Left Click Here

144

Chapter 3: Learning To Detail Part Drawings

46. After selecting **Chamfer,** place a dot (left click) next to the text "Distance distance". Place another dot (left click) next to the text "Full preview" as shown in Figure 41.

Figure 41

Left Click Here

Chapter 3: Learning To Detail Part Drawings

47. Move the cursor to the front upper corner. A red line will appear as shown in Figure 42. Left click once.

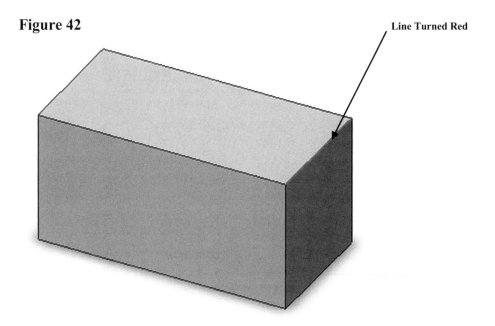

Figure 42 — Line Turned Red

48. A preview of the anticipated chamfer will be displayed as shown in Figure 43.

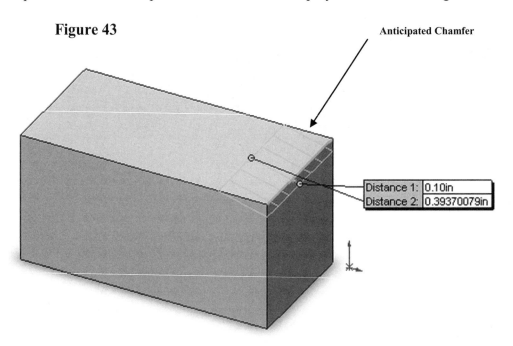

Figure 43 — Anticipated Chamfer

Distance 1: 0.10in
Distance 2: 0.39370079in

Chapter 3: Learning To Detail Part Drawings

49. Move the cursor to D1 in the Chamfer dialog box and highlight the text. Enter **.50**. Move the cursor to D2 in the dialog box and highlight the text. Enter **.75** as shown in Figure 44.

Figure 44

Chapter 3: Learning To Detail Part Drawings

50. A preview of the chamfer will be displayed as shown in Figure 45.

Figure 45

51. Left click on the green checkmark as shown in Figure 46.

Figure 46   Left Click Here

52. Your screen should look similar to Figure 47.

Figure 47

53. Move the cursor to the upper middle portion of the screen and left click on the "Rotate" icon as shown in Figure 48.

Figure 48 Left Click Here

54. While holding the left mouse button down, drag the cursor to the right to gain access to the backside of the part as shown in Figure 49.

Figure 49

Chapter 3: Learning To Detail Part Drawings

55. Right click anywhere on the screen. A pop menu will appear. Left click on **Select** as shown in Figure 50.

Figure 50

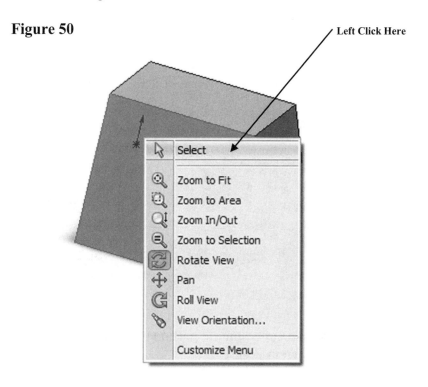

56. Move the cursor to the upper middle portion of the screen and left click on the drop down arrow next to the "Views Orientation" icon. A drop down menu will appear. Left click on **Left** as shown in Figure 51.

Figure 51

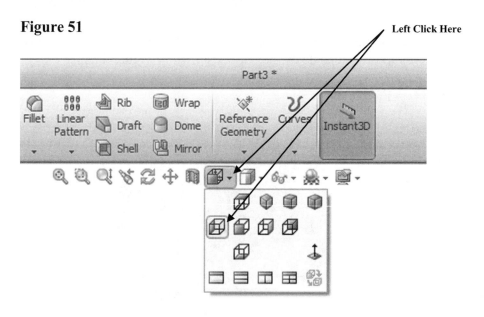

150

Chapter 3: Learning To Detail Part Drawings

57. The part will rotate and provide a perpendicular view of the surface as shown in Figure 52.

Figure 52

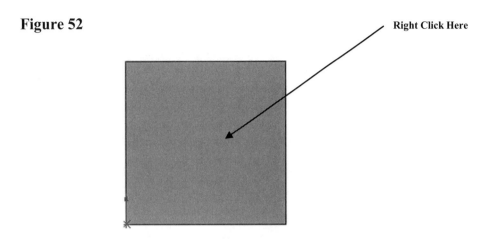

58. Right click anywhere on the surface. A pop up menu will appear. Left click on **Insert Sketch** as shown in Figure 53.

Figure 53

151

Chapter 3: Learning To Detail Part Drawings

59. A new sketch will appear on the surface as shown in Figure 54.

Figure 54

60. Move the cursor to the upper left portion of the screen and left click on **Circle** as shown in Figure 55.

Figure 55

61. Left click near the center of the part on the backside surface as shown in Figure 56.

Figure 56

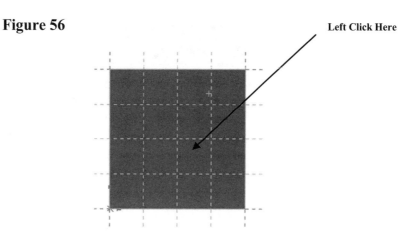

152

Chapter 3: Learning To Detail Part Drawings

62. Move the cursor to the side forming a circle and left click as shown in Figure 57.

Figure 57

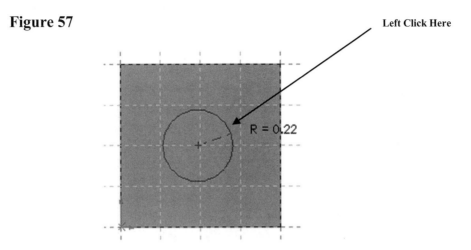

63. Right click anywhere on the screen. A pop up menu will appear. Left click on **Select** as shown in Figure 58.

Figure 58

64. Move the cursor to the upper left portion of the screen and left click on **Smart Dimension** as shown in Figure 59.

Figure 59

153

Chapter 3: Learning To Detail Part Drawings

65. After selecting **Smart Dimension** move the cursor over the edge of the circle until it turns red as shown in Figure 60. Left click on the edge of the circle.

Figure 60

66. Move the cursor to where the dimension will be placed and left click once as shown in Figure 61.

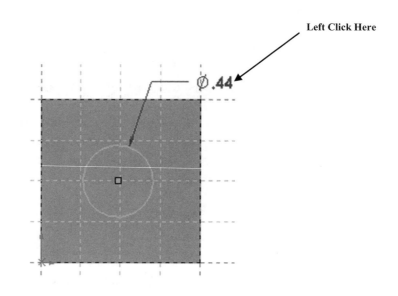

Figure 61

154

Chapter 3: Learning To Detail Part Drawings

67. The Modify dialog box will appear as shown in Figure 62.

Figure 62

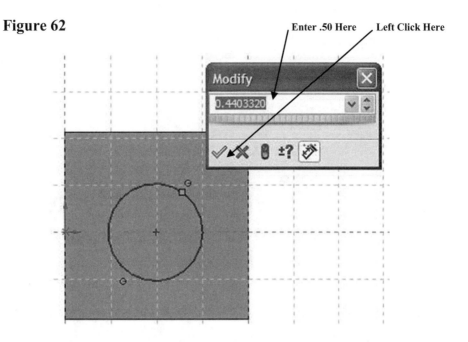

68. To edit the dimension, enter **.50** in the Modify dialog box (while the current dimension is highlighted) and either press **Enter** on the keyboard or left click on the green checkmark as shown in Figure 62.

69. Your screen should look similar to Figure 63.

Figure 63

Chapter 3: Learning To Detail Part Drawings

70. Move the cursor over the edge of the part until it turns red. Select the edge by left clicking anywhere on the edge as shown in Figure 64.

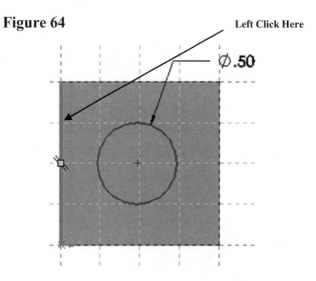

Figure 64

71. Move the cursor over the center of the circle until it turns red and left click as shown in Figure 65.

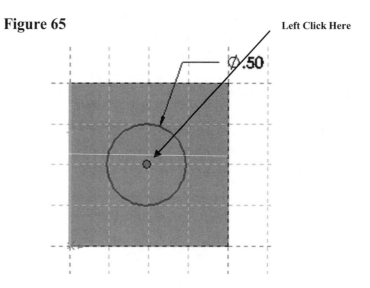

Figure 65

72. Move the cursor to where the dimension will be placed and left click once as shown in Figure 66.

Figure 66

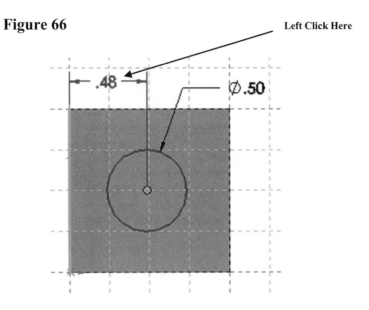

73. The Modify dialog box will appear as shown in Figure 67.

Figure 67

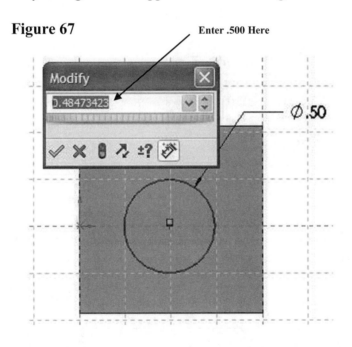

Chapter 3: Learning To Detail Part Drawings

74. To edit the dimension, enter **.500** in the Modify dialog box (while the current dimension is highlighted) and either press **Enter** on the keyboard or left click on the green checkmark as shown in Figure 68.

Figure 68

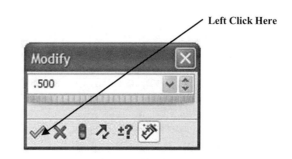

75. Your screen should look similar to Figure 69.

Figure 69

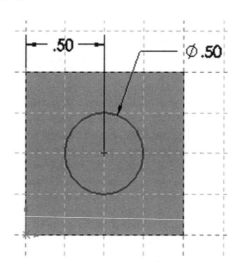

76. Move the cursor to the upper left portion of the screen and left click on **Smart Dimension** as shown in Figure 70.

Figure 70

158

77. Move the cursor to the edge of the part until it turns red. Left click anywhere on the edge of the part as shown in Figure 71.

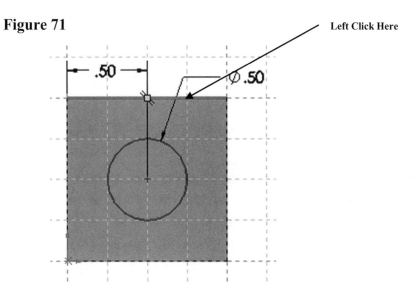

Figure 71

78. Move the cursor over the center of the circle until it turns red and left click as shown in Figure 72.

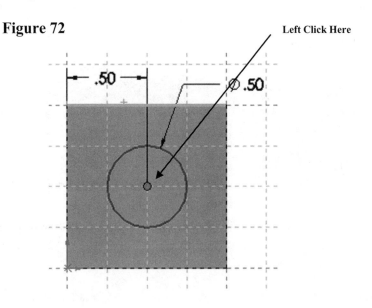

Figure 72

79. Move the cursor to where the dimension will be placed and left click once as shown in Figure 73.

Figure 73

80. The Modify dialog box will appear as shown in Figure 74.

Figure 74

81. To edit the dimension, enter **.500** in the Modify dialog box (while the current dimension is highlighted) and either press **Enter** on the keyboard or left click on the green checkmark as shown in Figure 75.

Figure 75

82. Your screen should look similar to Figure 76.

Figure 76

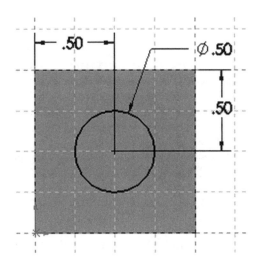

83. After the sketch is complete it is time to extrude a hole through the solid.

84. Right click anywhere on the screen. A pop up menu will appear. Left click on **Select** as shown in Figure 77. This will ensure that no commands are active.

Figure 77

161

Chapter 3: Learning To Detail Part Drawings

85. After you have verified that no commands are active, move the cursor to the upper left portion of the screen and left click on **Exit Sketch** as shown in Figure 78.

Figure 78

86. SolidWorks is now out of the Sketch commands and into the Features commands. Notice that the commands at the top of the screen are now different. To work in the Features commands a sketch must be present and have no opens (non-connected lines). If there are any opens in the sketch an error message will appear. Your screen should look similar to Figure 79.

Figure 79

87. Move the cursor to the upper middle portion of the screen and left click on the "Rotate" icon as shown in Figure 80.

Figure 80

162

Chapter 3: Learning To Detail Part Drawings

88. While holding the left mouse button down, drag the cursor to the right to gain an isometric view of the part as shown in Figure 81.

Figure 81

89. Right click anywhere on the screen. A pop up menu will appear. Left click on **Select** as shown in Figure 82.

Figure 82

Left Click Here

Chapter 3: Learning To Detail Part Drawings

90. Move the cursor to the upper left portion of the screen and left click on **Extruded Cut** as shown in Figure 83. If you received an error message, there are opens (non-connected lines) somewhere on the sketch. Check each intersection for opens by using the **Extend** and **Trim** commands.

Figure 83

91. Left click on the edge of the circle as shown in Figure 84.

Figure 84

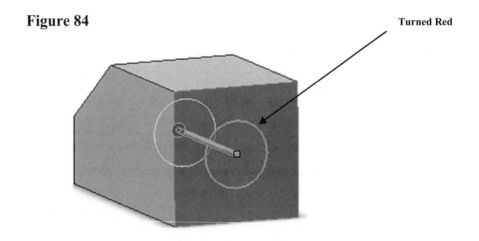

Chapter 3: Learning To Detail Part Drawings

92. Enter **2.00** for D1 as shown in Figure 85.

Figure 85

93. A preview will be displayed of the extrusion as shown in Figure 86.

Figure 86

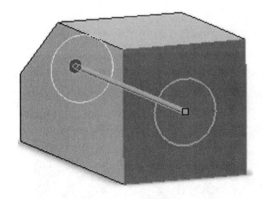

165

94. Left click on the green checkmark as shown in Figure 87.

Figure 87

95. Move the cursor to the upper middle portion of the screen and left click on the drop down arrow next to the "Views Orientation" icon. A drop down menu will appear. Left click on **Trimetric** as shown in Figure 88.

Figure 88

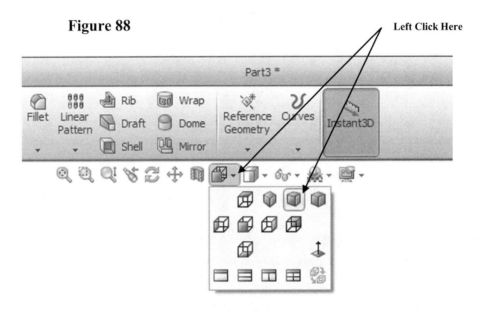

96. Your screen should look similar to Figure 89.

Figure 89

Chapter 3: Learning To Detail Part Drawings

97. Save the part file for easy retrieval to be used in the following section.

98. After the part file has been saved, move the cursor to the upper left portion of the screen and left click on the drop down arrow to the right of the "New Drawing" icon. A drop down menu will appear. Left click on **Make Drawing from Part/Assembly** as shown in Figure 90.

Figure 90

99. The Sheet Format/Size dialog box will open as shown in Figure 91.

Figure 91

100. Left click on **A-Landscape** as shown in Figure 91.

101. Left click on **OK**.

167

102. Drag any view, that you want to be the Front view, off of the View Palette at the right (not shown) on to the sheet. After placing the Front view, move the cursor up and left click once. The Top view will be attached to the cursor and will appear. Continue placing the Side view which will also be attached to the cursor. After the views have been placed, move the cursor to the upper right portion of the screen and right click on **Sheet Format**. A pop up menu will appear. Left click on **Properties** as shown in Figure 92.

Figure 92

103. The Sheet Properties dialog box will appear. You may want to increase/decrease the scale to fit the work areas. Left click next to Third Angle Projection. Left click on **OK** as shown in Figure 93.

Figure 93

168

Chapter 3: Learning To Detail Part Drawings

104. The following tools should be visible. If the tools are not visible, left click on the **View Layout** tab as shown in Figure 94.

Figure 94

105. **NOTE: If you were able to create a 3 view drawing buy dragging views off the View Pallet then skip to step 107.** Move the cursor to the upper left portion of the screen and left click on **Standard 3 View** as shown in Figure 95.

Figure 95

106. If the file name you have been working on appears as shown in Figure 96, move the cursor over the file name and double click. If the file name does not appear, use the Browse button to locate the file. SolidWorks will create a 3 view drawing.

Figure 96

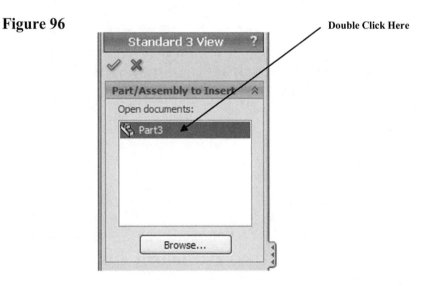

Chapter 3: Learning To Detail Part Drawings

107. Move the cursor to the upper left portion of the screen and left click on **Projected View** as shown in Figure 97.

Figure 97

108. Move the cursor to the Front View causing a red dashed box to appear. Left click inside the box as shown in Figure 98.

Figure 98

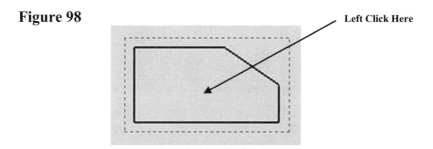

109. Move the cursor to the upper right portion of the screen. An isometric view of the part will be attached to the cursor. Left click as shown in Figure 99.

Figure 99

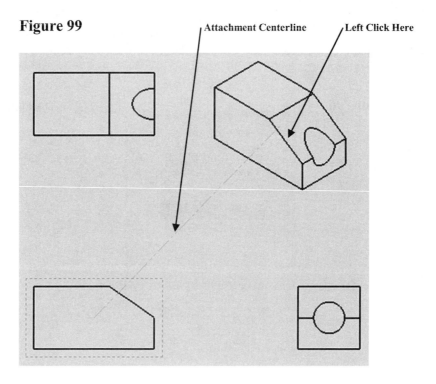

Chapter 3: Learning To Detail Part Drawings

110. Move the cursor to the Front View causing a red dashed box to appear. Left click inside the box as shown in Figure 100. The Feature Manager will appear to the left.

Figure 100

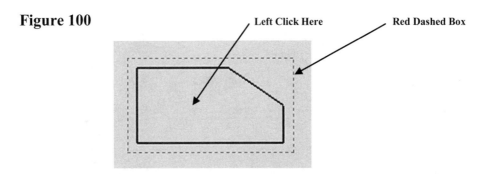

111. Move the cursor to the middle left portion of the screen under "Display Style" and left click on the "Hidden Lines Visible" icon as shown in Figure 101.

Figure 101

112. Move the cursor to the upper left portion of the screen and left click on the green checkmark as shown in Figure 102.

Figure 102

171

Chapter 3: Learning To Detail Part Drawings

113. Your screen should look similar to Figure 103.

Figure 103

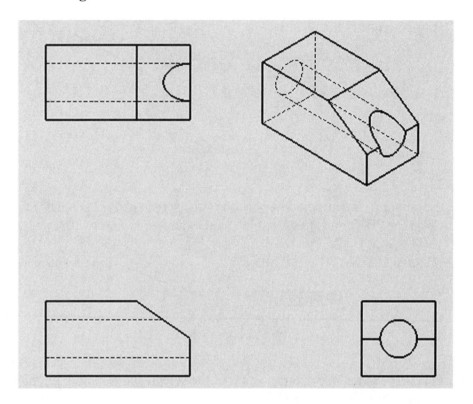

114. If for some reason the views need to be relocated, (moved closer together for example) move the cursor to any of the views (the top view in this case) causing red dots to appear and left click on the red dashed line (holding the left mouse button down). Drag the cursor down to move the view to the desired location as shown in Figure 104.

Figure 104 Left Click Here, Hold Down and Drag

115. Once the views have been moved closer together move the cursor to the upper left portion of the screen and left click on the green checkmark as shown in Figure 105.

Figure 105

116. Your screen may look similar to Figure 106.

Figure 106

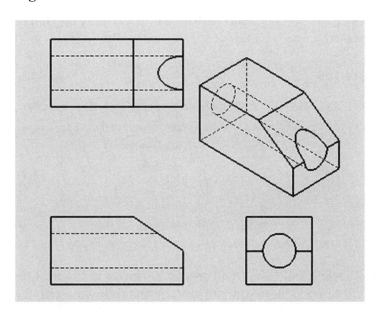

117. Move the cursor to the isometric view causing a red dashed box to appear around the part. Left click once. Move the cursor to the middle left portion of the screen and left click on the "Shaded with Edges" icon as shown in Figure 107.

Figure 107

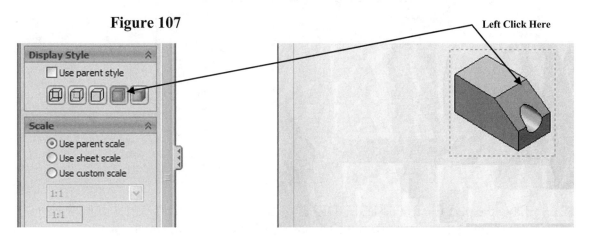

173

Chapter 3: Learning To Detail Part Drawings

118. Once you have created the isometric view, it will need to be deleted in order to create more room on the drawing. Move the cursor around the isometric view causing a red dashed box to appear around the part as shown in Figure 108.

Figure 108

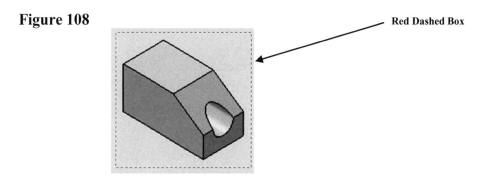

Red Dashed Box

119. Right click once. A pop up menu will appear. Left click on **Delete** as shown in Figure 109.

Figure 109

Left Click Here

120 The Confirm Delete dialog box will appear as shown in Figure 110. Left click on **Yes**. This will provide more room to work in the following chapter.

Figure 110

Left Click Here

121. Save the part file as Part1.SLDDRW for easy retrieval. This part will be used in the following chapter. **Be sure to close SolidWorks completely out**.

174

Chapter 3: Learning To Detail Part Drawings

Drawing Activities

Use these problems from Chapters 1 and 2 to create 3 view orthographic view detail drawings.

Problem 1

Problem 2

175

Chapter 3: Learning To Detail Part Drawings

Problem 3

Problem 4

Problem 5

Problem 6

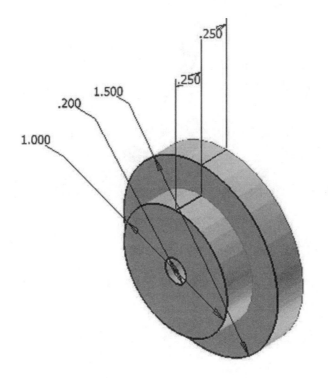

Chapter 3: Learning To Detail Part Drawings

Problem 7

Problem 8

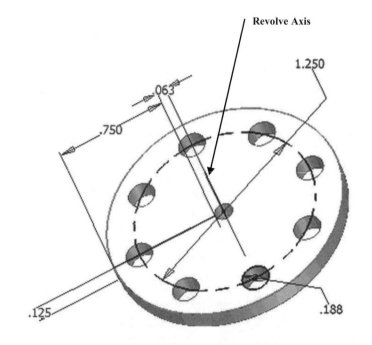

Chapter 4 Advanced Detail Drawing Procedures

Objectives:
- Create an Auxiliary View using the Auxiliary View command
- Dimension views using the Smart Dimension command
- Create a Section View using the Section View command
- Create Text using the Note command

Chapter 4 includes instruction on how to create the drawings shown below.

Chapter 4: Advanced Detail Drawing Procedures

1. Start SolidWorks by referring to "Chapter 1 Getting Started". SolidWorks will have to be started up from "scratch" before you begin this chapter.

2. After SolidWorks is running, open the .SLDDRW file that was created in Chapter 3. Move the cursor to the upper left corner of the screen and left click on "Open" as shown in Figure 1.

Figure 1

3. The Open dialog box will appear. Left click on the drawing that was created in Chapter 3. Make sure to select .slddrw from the file extension drop down menu as shown in Figure 2.

Figure 2

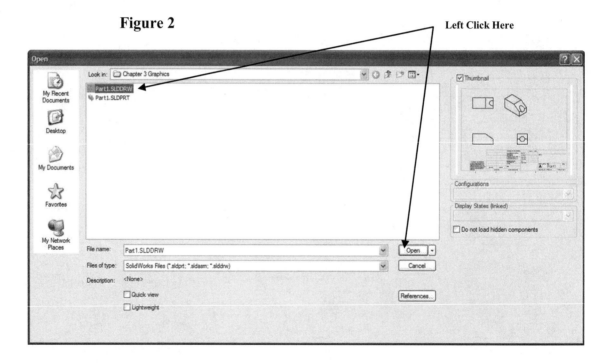

4. Left Click on **Open**.

180

Chapter 4: Advanced Detail Drawing Procedures

5. After the .SLDDRW file is open, move the views closer to each other to provide additional room on the drawing. Start by moving the cursor over the top view. Dots will appear around the view. Left click (holding the left mouse button down) on the dots and drag the view down closer to the front view as shown in Figure 3.

Figure 3 — Left Click Here, Hold Down and Drag

6. Move the side view closer to the front view. Start by moving the cursor over the side view. Dots will appear around the view. Left click (holding the left mouse button down) on the dots and drag the view closer to the front view as shown in Figure 4.

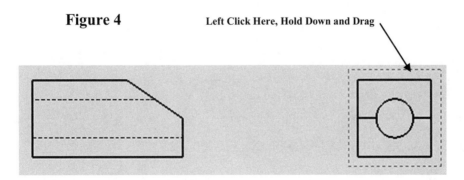

Figure 4 — Left Click Here, Hold Down and Drag

Chapter 4: Advanced Detail Drawing Procedures

7. This will provide more room to work. Your screen should look similar to Figure 5. Press **ESC** on the keyboard.

Figure 5

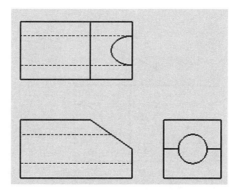

8. Move the cursor to the upper left portion of the screen and left click on the **View Layout** tab as shown in Figure 6.

Figure 6

9. Notice that the commands at the top of the screen are now different as shown in Figure 7.

Figure 7

10. Left click on **Auxiliary View** as shown in Figure 8.

Figure 8

11. Move the cursor to the wedge line and left click as shown in Figure 9.

Figure 9

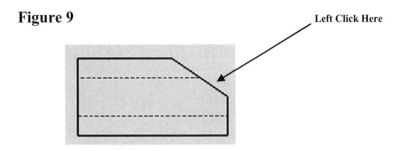

Left Click Here

12. SolidWorks will create an auxiliary view from the selected surface. The view will be attached to the cursor as shown in Figure 10.

Figure 10

Auxiliary View

183

Chapter 4: Advanced Detail Drawing Procedures

13. Drag the cursor towards the other views and left click as shown in Figure 11.

Figure 11

14. Move the cursor over to the upper right portion of the screen and left click on the green checkmark as shown in Figure 12.

Figure 12

Chapter 4: Advanced Detail Drawing Procedures

15. Your screen should look similar to Figure 13.

Figure 13

16. Move the cursor to the side view. Red dots will appear as shown in Figure 14.

Figure 14

Chapter 4: Advanced Detail Drawing Procedures

17. Right click on the view. A pop up menu will appear. Left click on **Delete** as shown in Figure 15.

Figure 15

18. The Confirm Delete dialog box will appear. Left click on **Yes** as shown in Figure 16.

Figure 16

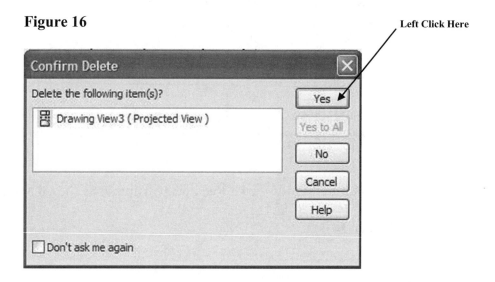

19. Move the cursor to the upper middle portion of the screen and left click on **Section View** as shown in Figure 17.

Figure 17

20. Move the cursor over the front view. Red dots will appear around the view. Left click above the top of the part as shown in Figure 18.

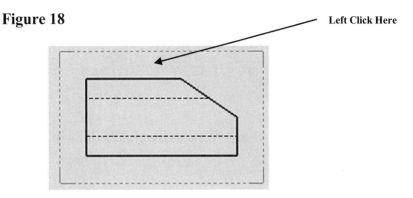

Figure 18 — Left Click Here

21. Drag the cursor down across the part and left click as shown in Figure 19.

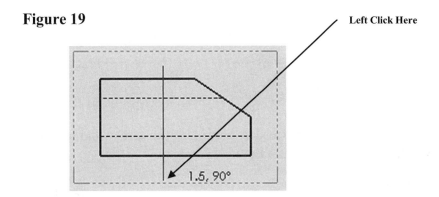

Figure 19 — Left Click Here

22. Drag the cursor to the right. The section view will be attached to the cursor. Place a check in the "Flip Direction" box if cutting plane arrows are pointing the wrong direction. Place the section view to the side of the front view and left click as shown in Figure 20.

Figure 20 — Left Click Here

Chapter 4: Advanced Detail Drawing Procedures

23. Move the cursor to the upper left portion of the screen and left click on the green checkmark as shown in Figure 21.

Figure 21

24. SolidWorks will create a section view to the right as shown in Figure 22.

Figure 22

25. Move the cursor to the upper left portion of the screen and left click on the **Annotation** tab as shown in Figure 23.

Figure 23

26. Notice that the commands at the top of the screen are now different as shown in Figure 24.

Figure 24

188

Chapter 4: Advanced Detail Drawing Procedures

27. Annotations commands are typically where "drafting" activities are performed.

28. Move the cursor to the upper left portion of the screen and left click on **Smart Dimension** as shown in Figure 25.

Figure 25

29. Move the cursor over the left side vertical line until it turns red as shown in Figure 26. Select the line by left clicking anywhere on the line **or** on each of the end points. To use the end points of the line, move the cursor over one of the end points. A small red dot will appear. Left click once and move the cursor to the other end point. After the red dot appears, left click once. The dimension will be attached to the cursor.

Figure 26

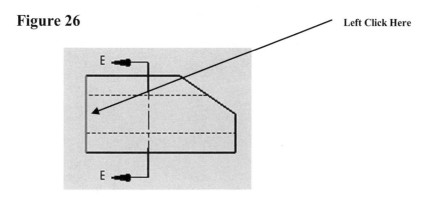

189

Chapter 4: Advanced Detail Drawing Procedures

30. Move the cursor to where the dimension will be placed and left click once as shown in Figure 27.

Figure 27

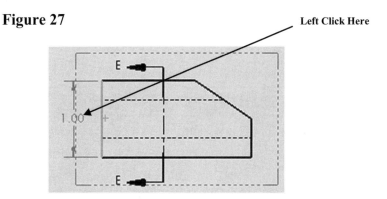

Left Click Here

31. Move the cursor to the upper left portion of the screen and left click on the green checkmark as shown in Figure 28.

Figure 28

Left Click Here

190

32. Your screen should look similar to Figure 29.

Figure 29

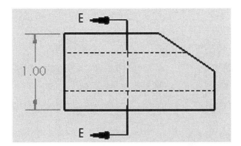

33. Finish dimensioning the part to your own satisfaction. When the part is satisfactorily dimensioned, save the file to a location where it can be easily retrieved.

34. **If** there is ever a need to hide any dimension, move the cursor over the dimension. The dimension will turn red as shown in Figure 30.

Figure 30

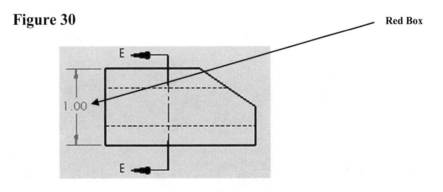

Chapter 4: Advanced Detail Drawing Procedures

35. Right click on the dimension. A pop up menu will appear. Left click on **Hide** as shown in Figure 31. To retrieve the dimension, immediately left click on the **Undo** icon at the far upper middle portion of the screen.

Figure 31

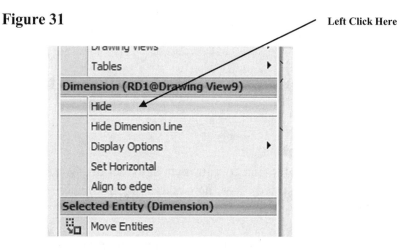

36. Left click on **Note** as shown in Figure 32.

Figure 32

Chapter 4: Advanced Detail Drawing Procedures

37. Move the cursor to the title block location. A small box will be attached to the cursor. Left click as shown in Figure 33.

Figure 33

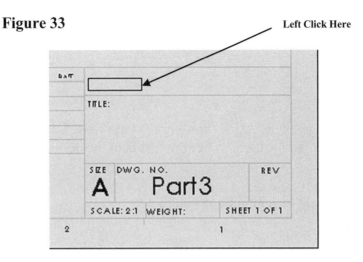

38. The cursor will begin blinking. Enter your first name and last name. After the text has been entered, move the cursor to the upper left portion of the screen and left click on the green checkmark as shown in Figure 34.

Figure 34

39. Text will appear in the title block as shown in Figure 35. Double clicking on the text will allow you to edit the text if needed.

Figure 35

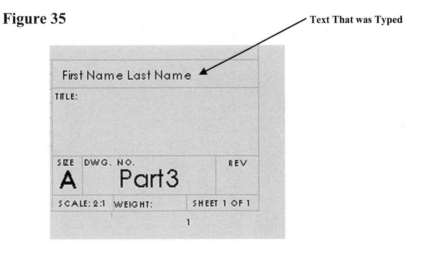

193

Chapter 4: Advanced Detail Drawing Procedures

40. If the text needs to be moved, move the cursor over the text causing it to turn green as shown in Figure 36.

Figure 36

41. While the text is green, left click (holding the left mouse button down) and drag the text to the desired location. After the text is in the desired location, release the left mouse button.

42. Your screen should look similar to Figure 37.

Figure 37

43. Save the current sheet where it can be easily retrieved.

Drawing Activities

Create Section View Drawings for the following:

Problem 1

Problem 2

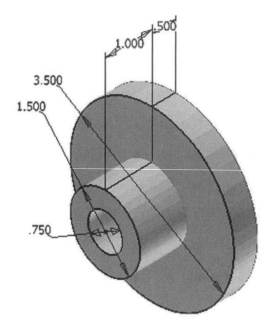

Create Auxiliary View Drawings for the following:

Problem 3

Problem 4

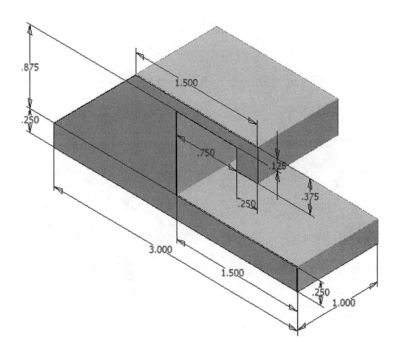

Chapter 4: Advanced Detail Drawing Procedures

Create Section View Drawings for the following:

Problem 5

Problem 6

Create Section View Drawings for the following:

Problem 7

Problem 8

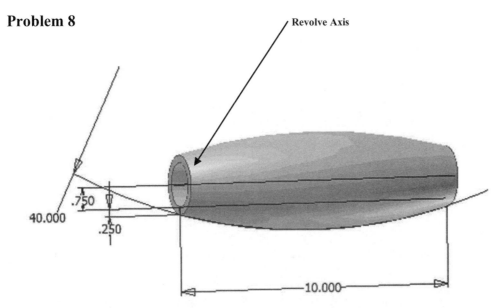

Chapter 5 Learning To Edit Existing Solid Models

Objectives:
- Design a simple part
- Learn to use the Circular Pattern command
- Learn to edit a part using the Circular Pattern command
- Edit the part using the Sketch command
- Edit the part using the Extruded Boss/Base command
- Edit the part using the Extruded Cut command
- Edit the part using the Chamfer command

Chapter 5 includes instruction on how to design and edit the part shown below.

Chapter 5: Learning To Edit Existing Solid Models

1. Start SolidWorks by referring to "Chapter 1 Getting Started".

2. After SolidWorks is running, begin a new sketch.

3. Move the cursor to the upper left corner of the screen and left click on **Circle** as shown in Figure 1.

Figure 1

4. Move the cursor to the center of the screen and left click once. This will be the center of the circle as shown in Figure 2.

Figure 2

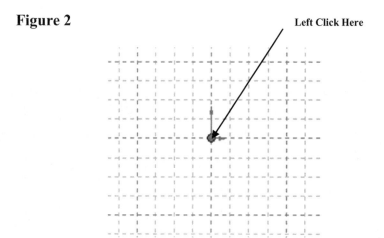

5. Move the cursor to the right and left click once as shown in Figure 3.

Figure 3

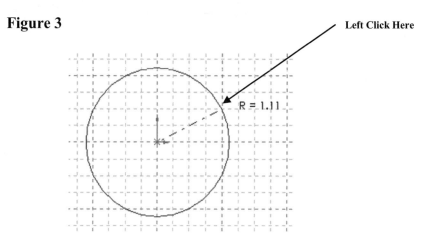

202

6. Right click anywhere on the drawing. A pop up menu will appear. Left click on **Select** as shown in Figure 4

Figure 4

7. Move the cursor to the middle left portion of the screen and left click on **Smart Dimension** as shown in Figure 5.

Figure 5

8. Move the cursor over the edge (not center) of the circle until it turns red. Left click once as shown in Figure 6. The dimension will be attached to the cursor.

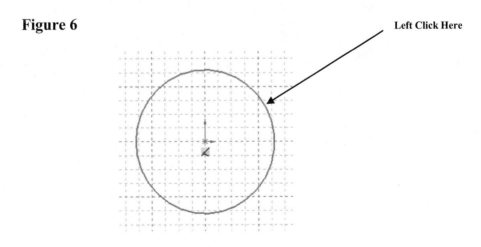

Figure 6

Chapter 5: Learning To Edit Existing Solid Models

9. Move the cursor to where the dimension will be placed and left click once as shown in Figure 7.

Figure 7

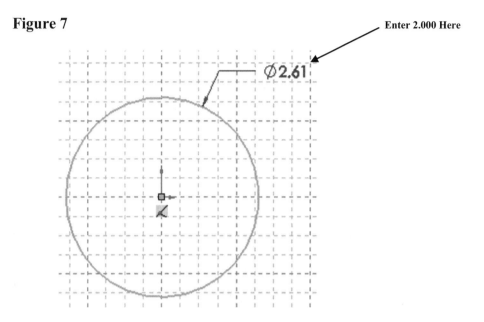

10. The Modify dialog box will appear as shown in Figure 8.

Figure 8

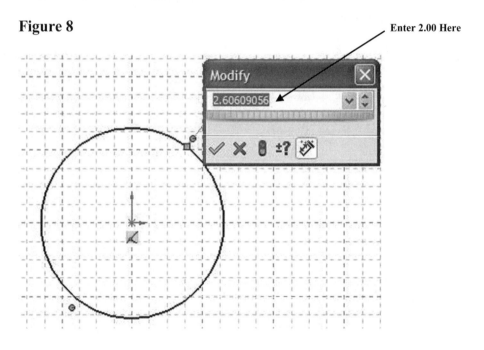

204

11. To edit the dimension, enter **2.00** in the Modify dialog box (while the current dimension is highlighted) and left click on the green checkmark as shown in Figure 9.

Figure 9   **Left Click Here**

12. The dimension of the circle will become 2.00 inches as shown in Figure 10.

Figure 10   **Left Click Here**

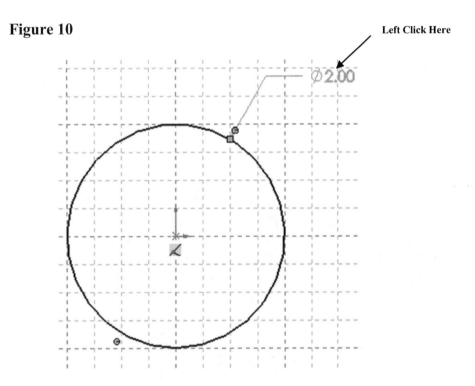

Chapter 5: Learning To Edit Existing Solid Models

13. Right click anywhere around the drawing. A pop up menu will appear. Left click on **Select** as shown in Figure 11.

Figure 11

14. To view the entire drawing move the cursor to the middle portion of the screen and left click on the "Zoom To Fit" icon as shown in Figure 12.

Figure 12

15. The drawing will "fill up" the entire screen. If the drawing is still too large, left click on the "Zoom In/Out" icon as shown in Figure 13. After selecting the "Zoom In/Out icon, hold the left mouse button down and drag the cursor up and down to achieve the desired view of the sketch.

Figure 13

Chapter 5: Learning To Edit Existing Solid Models

16. After the sketch is complete it is time to extrude the sketch into a solid. Right click anywhere on the drawing. A pop up menu will appear. Left click on **Select** as shown in Figure 14.

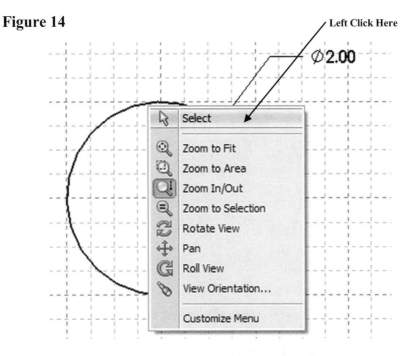

Figure 14

17. After you have verified that no commands are active, right click anywhere on the sketch. A pop up menu will appear. Left click on the **Exit Sketch** icon as shown in Figure 15. If the Features tools did not appear, left click on the **Features** tab at the upper left portion of the screen.

Figure 15

207

18. SolidWorks is now out of the Sketch commands and into the Features commands. Notice that the commands at the top of the screen and now different. To work in the Features commands a sketch must be present and have no opens (non-connected lines). If there are any opens in the sketch an error message will appear. You screen should look similar to Figure 16.

Figure 16

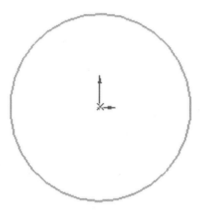

19. Move the cursor to the upper middle portion of the screen and left click on the drop down arrow next to the "Views Orientation" icon. A drop down menu will appear. Left click on **Trimetric** as shown in Figure 17.

Figure 17

Left Click Here

20. The view will become trimetric as shown in Figure 18.

Figure 18

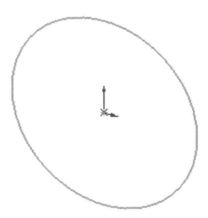

21. Move the cursor to the upper left portion of the screen and left click on **Extruded Boss/Base**. If the **Extruded Boss/Base** command does not appear, left click on the **Features** tab as shown in Figure 19.

Figure 19

Left Click Here

22. A preview of the extrusion will be displayed. <u>If a preview is not displayed, left click anywhere on the edge of the circle.</u> If SolidWorks gave you an error message, there are opens (non-connected lines) somewhere on the sketch. Check each intersection for opens by using the **Extend** and **Trim** commands.

Chapter 5: Learning To Edit Existing Solid Models

23. Move the cursor to the upper left portion of the screen and enter **.25** next to D1 as shown in Figure 20.

Figure 20

Enter .25 Here

24. Move the cursor to the upper left portion of the screen and left click on the green checkmark as shown in Figure 21.

Figure 21

Left Click Here

Chapter 5: Learning To Edit Existing Solid Models

25. Your screen should look similar to Figure 22.

   Figure 22

26. Move the cursor to the front surface causing the edges of the surface to turn red. Right click on the surface once. A pop up menu will appear. Left click on the **Insert Sketch** icon as shown in Figure 23.

   Figure 23

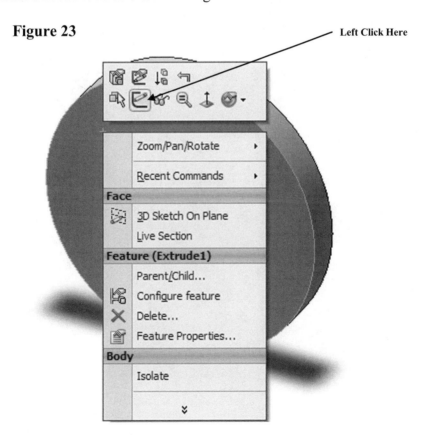

211

Chapter 5: Learning To Edit Existing Solid Models

27. SolidWorks will start a new sketch on the selected surface as shown in Figure 24.

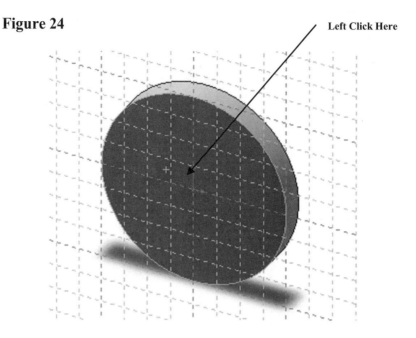

Figure 24

Left Click Here

28. Move the cursor to the upper right portion of the screen and left click on the drop down arrow next to the "Standard Views" icon. A drop down menu will appear. Left click on the **Front** icon. You can also left click on the **Normal To** icon as shown in Figure 25.

Figure 25

Left Click Here

Chapter 5: Learning To Edit Existing Solid Models

29. The part will rotate providing a perpendicular view of the surface as shown in Figure 26.

   Figure 26

   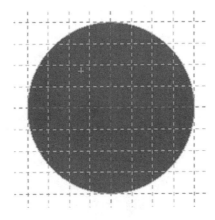

30. Move the cursor to the upper middle portion of the screen and left click on **Circle** as shown in Figure 27.

   Figure 27   Left Click Here

31. Move the cursor to the center of the circle. A red dot will appear in the center as shown in Figure 28.

   Figure 28   Turned Red

   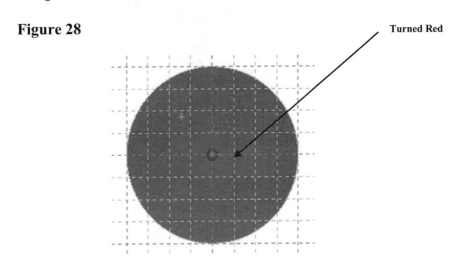

213

Chapter 5: Learning To Edit Existing Solid Models

32. After the red dot appears, move the cursor straight up causing a blue dashed line to appear. Left click once as shown in Figure 29.

Figure 29

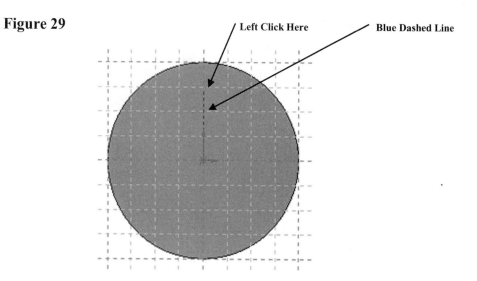

33. Move the cursor to the right side. A circle will form. Let click once as shown in Figure 30.

Figure 30

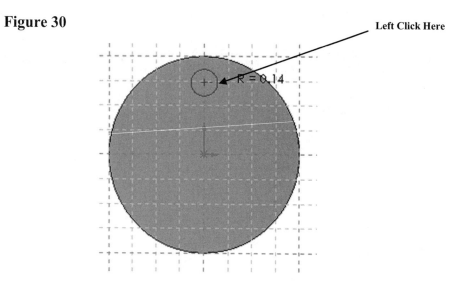

Chapter 5: Learning To Edit Existing Solid Models

34. Right click anywhere on the drawing. A pop up menu will appear. Left click on **Select** as shown in Figure 31.

Figure 31

35. Move the cursor to the upper left portion of the screen and left click on **Smart Dimension** as shown in Figure 32.

Figure 32

36. Move the cursor over the edge (not center) of the circle until it turns red as shown in Figure 33. Select the circle by left clicking anywhere on the edge. The dimension will be attached to the cursor.

Figure 33

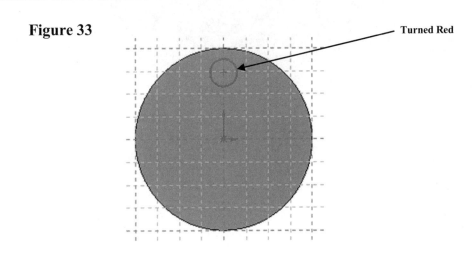

215

Chapter 5: Learning To Edit Existing Solid Models

37. Move the cursor to where the dimension will be placed and left click once as shown in Figure 34.

Figure 34

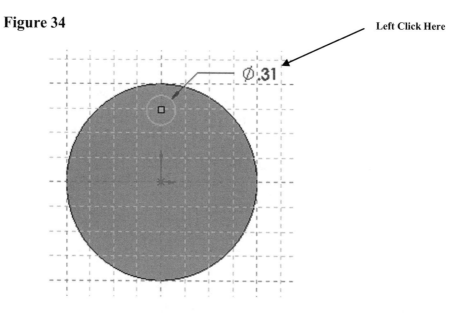

38. The Modify dialog box will appear as shown in Figure 35.

Figure 35

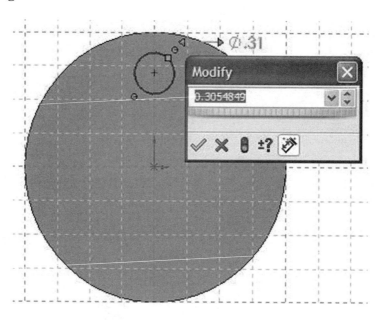

216

39. To edit the dimension, type **.375** in the Modify dialog box (while the current dimension is highlighted) and left click on the green check mark as shown in Figure 36.

Figure 36

40. The dimension of the circle will become .375 inches as shown in Figure 37. Use the Zoom icons to zoom out if necessary.

Figure 37

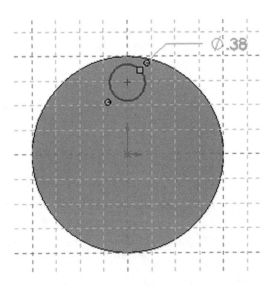

Chapter 5: Learning To Edit Existing Solid Models

41. Right click anywhere around the drawing. A pop up menu will appear. Left click on **Select** as shown in Figure 38.

Figure 38

42. After you have verified that no commands are active, right click anywhere on the sketch. A pop up menu will appear. Left click on the **Exit Sketch** icon as shown in Figure 39.

Figure 39

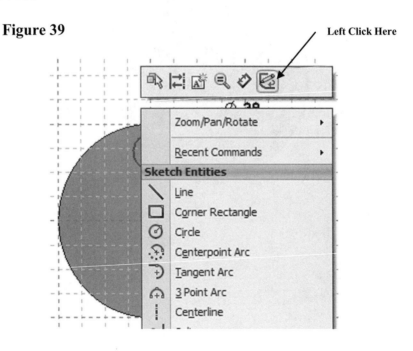

218

43. SolidWorks is now out of the Sketch commands and into the Features commands. Notice that the commands at the top of the screen are now different. To work in the Features commands a sketch must be present and have no opens (non-connected lines). If there are any opens in the sketch an error message will appear. Your screen should look similar to Figure 40.

Figure 40

44. Move the cursor to the upper right portion of the screen and left click on the drop down arrow next to the "Views Orientation" icon. A drop down menu will appear. Left click on **Trimetric** as shown in Figure 41.

Figure 41   Left Click Here

Chapter 5: Learning To Edit Existing Solid Models

45. The view will become trimetric as shown in Figure 42.

Figure 42

46. Move the cursor to the upper middle portion of the screen and left click on **Extruded Cut** as shown in Figure 43.

Figure 43

47. A preview of the cut extrusion will be displayed. If a preview is not displayed left click on the edge of the circle. SolidWorks gave you an error message, there are opens (non-connected lines) somewhere on the sketch. Check each intersection for opens by using the Extend and Trim commands. Your screen should look similar to Figure 44.

Figure 44

48. Enter **.25** next to D1 as shown in Figure 45.

Figure 45

Enter .25 Here

49. Move the cursor to the upper left portion of the screen and left click on the green checkmark as shown in Figure 46.

Figure 46

Left Click Here

Chapter 5: Learning To Edit Existing Solid Models

50. Your screen should look similar to Figure 47. You may have to use the zoom out command to view the part.

Figure 47

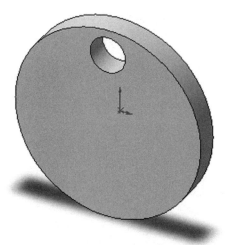

51. Move the cursor to the upper right portion of the screen and left click on the double arrow under Reference Geometry as shown in Figure 48.

Figure 48

Left Click Here

52. A drop down menu will appear. Left click on **Axis** as shown in Figure 49.

Figure 49

Left Click Here

222

Chapter 5: Learning To Edit Existing Solid Models

53. Move the cursor to the side of the part causing both edges to turn red. Left click once as shown in Figure 50.

Figure 50

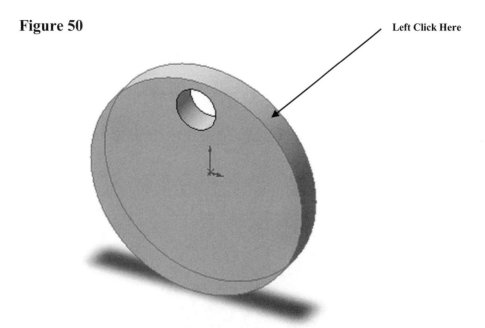

Left Click Here

54. SolidWorks will project a center axis as shown in Figure 51. This axis will be used when creating a circular pattern.

Figure 51

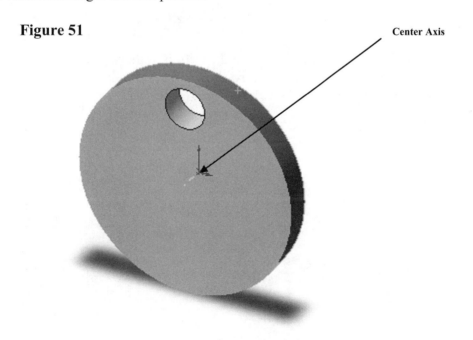

Center Axis

Chapter 5: Learning To Edit Existing Solid Models

55. Move the cursor to the middle portion of the screen and left click on the green checkmark as shown in Figure 52.

Figure 52

Left Click Here

56. The axis will be displayed as shown in Figure 53.

Figure 53

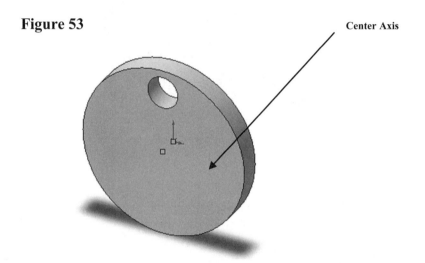

Center Axis

57. Move the cursor to the upper middle portion of the screen and left click on the drop down arrow under Linear Pattern. A drop down menu will appear. Left click on **Circular Pattern** as shown in Figure 54.

Figure 54

Left Click Here

Chapter 5: Learning To Edit Existing Solid Models

58. Move the cursor to the left portion of the screen and enter **3** for the number of holes and left click in the box next to the text "Equal spacing" as shown in Figure 55.

Figure 55

59. Move the cursor to the center of the circle (hole) as shown in Figure 56. The edges will become red lines. SolidWorks will only find the circle (hole) if the view is trimetric or isometric.

Figure 56

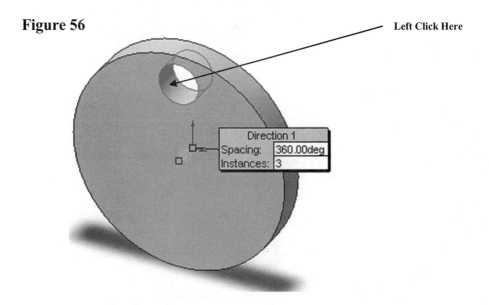

Chapter 5: Learning To Edit Existing Solid Models

60. Left click inside the circle (hole). The inside of the hole will turn green. A preview of the holes will be displayed as shown in Figure 57.

Figure 57

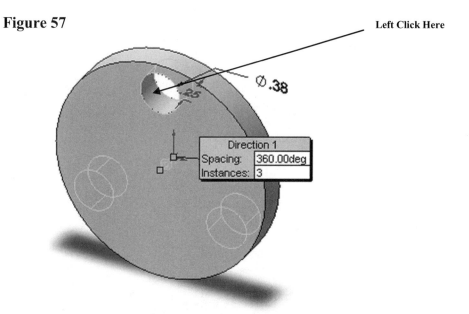

61. Move the cursor to the left portion of the screen and left click on the green checkmark as shown in Figure 58.

Figure 58

62. Your screen should look similar to Figure 59.

Figure 59

Chapter 5: Learning To Edit Existing Solid Models

63. Move the cursor to the upper middle portion of the screen and left click the drop down arrow under Fillet. A drop menu will appear. Left click on **Chamfer** as shown in Figure 60.

Figure 60

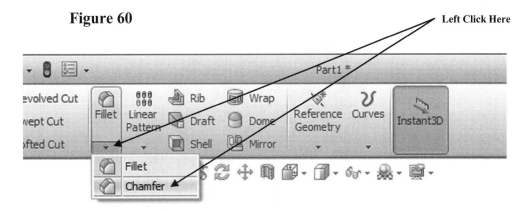

64. Enter **.0625** for the distance as shown in Figure 61. Left click next to Full Preview. Press **Enter** on the keyboard.

Figure 61

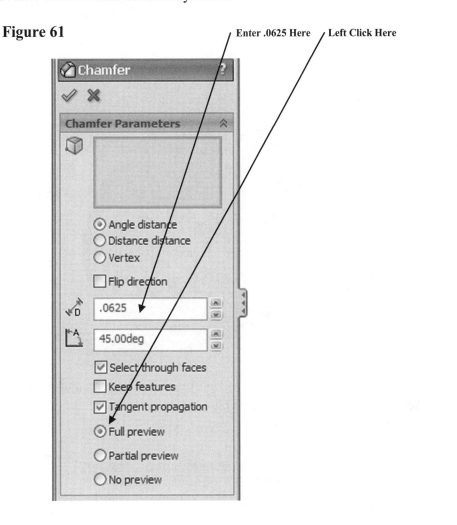

227

Chapter 5: Learning To Edit Existing Solid Models

65. Move the cursor over the edge of the part causing it to turn red as shown in Figure 62. Left click once.

Figure 62

66. A preview of the chamfer will be displayed as in Figure 63.

Figure 63

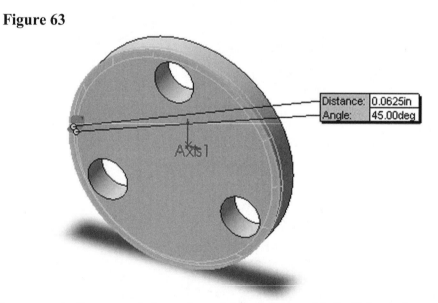

67. Move the cursor to the upper left portion of the screen and left click on the green checkmark as shown in Figure 64.

Figure 64

68. Your screen should look similar to Figure 65.

Figure 65

69. If for some reason a change needs to be made to this part, it can be accomplished by either editing a sketch or feature located in the Part Tree at the middle left corner of the screen as shown in Figure 66.

Figure 66

Chapter 5: Learning To Edit Existing Solid Models

70. The Part Tree is shown in Figure 67. Left click on each of the "plus" signs in the part tree. The tree will expand showing more details for part construction.

Figure 67

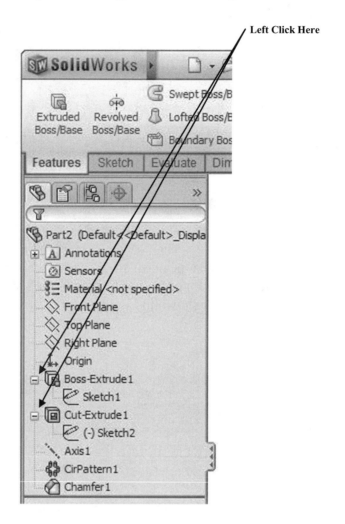

Chapter 5: Learning To Edit Existing Solid Models

71. Move the cursor over the text "Sketch1". The text will become highlighted as shown in Figure 68.

Figure 68

Highlighted Text

72. The original sketch will also appear as shown in Figure 69.

Figure 69

Original Sketch

Chapter 5: Learning To Edit Existing Solid Models

73. Right click on **Sketch1**. A pop up menu will appear. Left click on the **Edit Sketch** icon as shown in Figure 70.

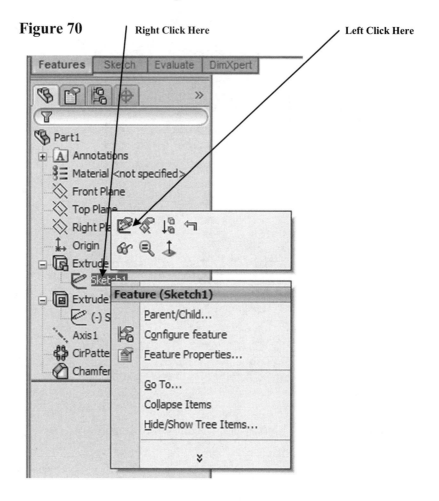

Figure 70

74. The original sketch will appear as shown in Figure 71.

Figure 71

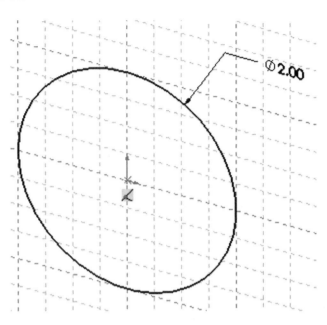

75. Move the cursor to the upper middle portion of the screen and left click on the drop down arrow next to the "Views Orientation" icon. A drop down menu will appear. Left click on **Front** or **Normal To** as shown in Figure 72.

Figure 72

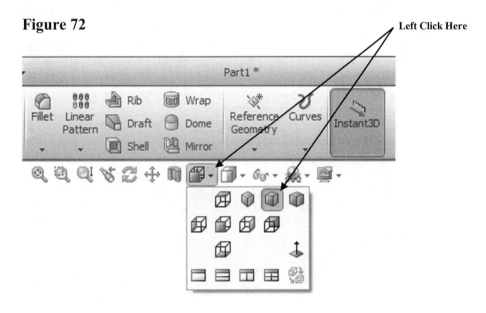

Left Click Here

Chapter 5: Learning To Edit Existing Solid Models

76. A perpendicular view of the sketch will be displayed. The sketch will look similar to when the sketch was first constructed. Your screen should look similar to Figure 73.

Figure 73

77. Start by modifying the diameter of the part. Double click on the overall dimension. The Modify dialog box will appear as shown in Figure 74.

Figure 74

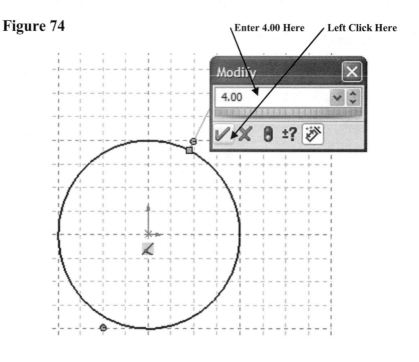

78. Enter **4.00** and left click on the green checkmark as shown in Figure 74.

79. The diameter of the part will increase to 4.00 as shown in Figure 75.

Figure 75

80. Move the cursor to the upper middle portion of the screen and left click on the green checkmark as shown in Figure 76.

Figure 76

Left Click Here

81. Right click anywhere on the screen. A pop up menu will appear. Left click on the **Exit Sketch** icon as shown in Figure 77.

Figure 77

Left Click Here

235

Chapter 5: Learning To Edit Existing Solid Models

82. SolidWorks will automatically update the part to reflect the changes made in the sketch. It will not be necessary to extrude the part again. Your screen should look similar to Figure 78.

Figure 78

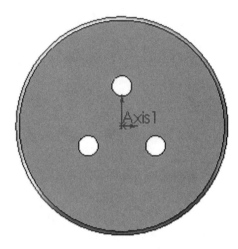

83. Move the cursor to the upper middle portion of the screen and left click on the drop down arrow next to the "View Orientation" icon. A drop down menu will appear. Left click on the **Trimetric** icon as shown in Figure 79.

Figure 79

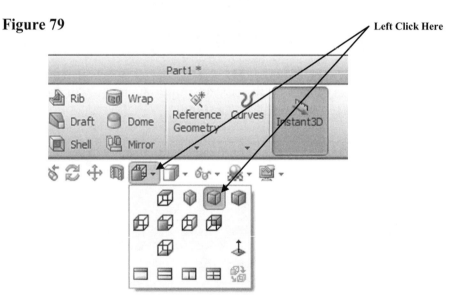

Left Click Here

Chapter 5: Learning To Edit Existing Solid Models

84. Your screen should look similar to Figure 80.

Figure 80

85. Move the cursor to the middle left portion of the screen where the part tree is located. Right click once on **Boss-Extrude1/Extrude1 (2009 and earlier versions of SolidWorks display "Extrude1" rather than "Boss-Extrude1"**. A pop up menu will appear. Left click on the **Edit Feature** icon as shown in Figure 81.

Figure 81

237

Chapter 5: Learning To Edit Existing Solid Models

86. Move the cursor to the middle left portion of the screen. The Extrude1 dialog box will appear. Enter **.750** for the extrusion distance and left click on the green checkmark as shown in Figure 82.

Figure 82

87. SolidWorks will automatically update the part without the need to repeat any of the steps that created the original part. Notice that the holes are no longer "thru" holes. Your screen should look similar to Figure 83.

Figure 83

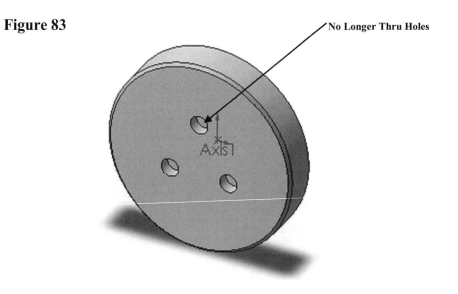

88. Move the cursor to the middle left portion of the screen where the part tree is located. Right click once on **Sketch2**. A pop up menu will appear. Left click on the **Edit Sketch** icon as shown in Figure 84.

Figure 84

89. The original sketch will appear as shown in Figure 85.

Figure 85

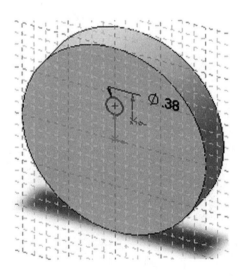

Chapter 5: Learning To Edit Existing Solid Models

90. Double click on the hole diameter dimension. The Modify dialog box will appear as shown in Figure 86.

Figure 86

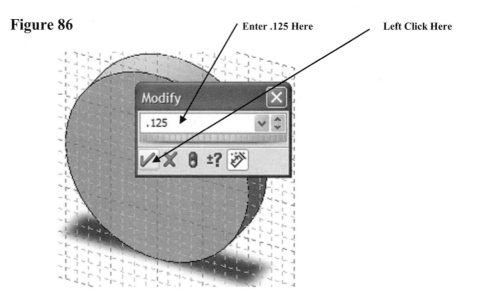

91. Enter **.125** and left click on the green checkmark as shown in Figure 86.

92. The diameter of the holes will be reduced to .125 as shown in Figure 87.

Figure 87

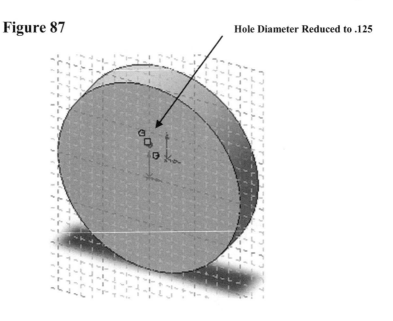

Chapter 5: Learning To Edit Existing Solid Models

93. Move the cursor to the upper left portion of the screen and left click on the green checkmark as shown in Figure 88.

Figure 88

Left Click Here

94. Right click anywhere around the drawing. A pop up menu will appear. Left click on the **Exit Sketch** icon as shown in Figure 89.

Figure 89

Left Click Here

Chapter 5: Learning To Edit Existing Solid Models

95. SolidWorks will automatically update the part as shown in Figure 90.

Figure 90

96. Move the cursor to the middle left portion of the screen where the part tree is located. Right click on **Cut-Extrude1/Extrude2 (2009 and earlier versions of SolidWorks display "Extrude2" rather than "Cut-Extrude1"**. A pop up menu will appear. Left click on the **Edit Feature** icon as shown in Figure 91.

Figure 91

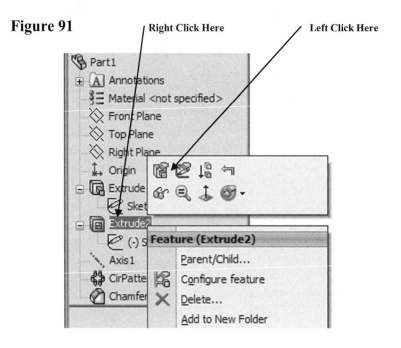

242

97. Enter **.750** of the extrusion distance as shown in Figure 92.

Figure 92

98. Move the cursor to the upper left portion of the screen and left click on the green checkmark as shown in Figure 93.

Figure 93

Chapter 5: Learning To Edit Existing Solid Models

99. SolidWorks will automatically update the part. Notice that the holes are now "thru" holes as shown in Figure 94.

Figure 94

100. Move the cursor to the upper middle portion of the screen and left click on the drop down arrow next to the "Views Orientation" icon. A drop down menu will appear. Left click on the **Front** icon as shown in Figure 95.

Figure 95

101. A perpendicular view of the part will be displayed. Verify the holes are actually thru holes as shown in Figure 96.

Figure 96

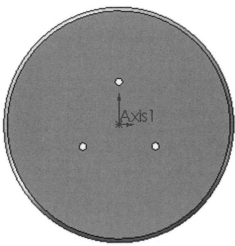

102. Move the cursor to the middle left portion of the screen where the part tree is located. Right click on **CirPattern1**. A pop up menu will appear. Left click on the **Edit Feature** icon as shown in Figure 97.

Figure 97

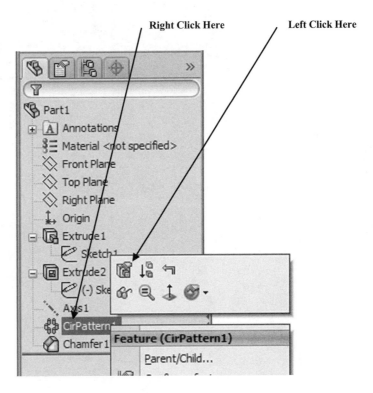

245

103. Enter **6** for the number of holes and left click next to Equal Spacing as shown in Figure 98. Press the **Enter** key on the keyboard.

Figure 98

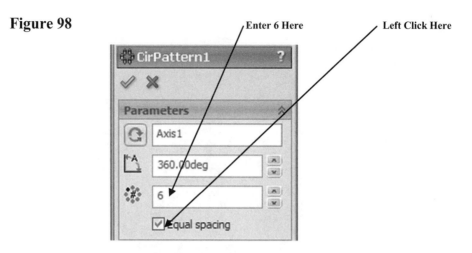

Enter 6 Here

Left Click Here

104. A preview will be displayed as shown in Figure 99.

Figure 99

105. Left click on the green checkmark as shown in Figure 100.

Figure 100

106. Your screen should look similar to Figure 101.

Figure 101

107. Move the cursor to the upper middle portion of the screen and left click on the drop down arrow next to the "Views Orientation" icon. A drop down menu will appear. Left click on the **Trimetric** icon as shown in Figure 102.

Figure 102

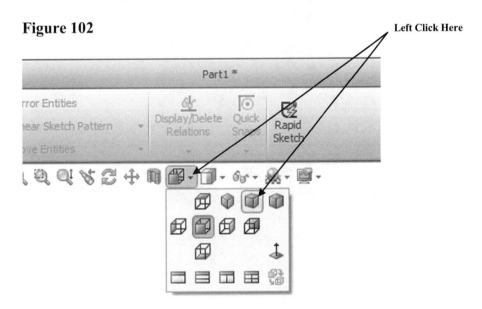

Chapter 5: Learning To Edit Existing Solid Models

108. The part will be displayed in trimetric as shown in Figure 103.

Figure 103

109. Move the cursor to the middle left portion of the screen where the part tree is located. Right click on **Chamfer1**. A pop up menu will appear. Left click on the **Edit Feature** icon as shown in Figure 104.

Figure 104

248

Chapter 5: Learning To Edit Existing Solid Models

110. Enter **.25** for the Distance a shown in Figure 105.

Figure 105

Enter .25 Here

111. A preview of the chamfer will be displayed as shown in Figure 106.

Figure 106

Chapter 5: Learning To Edit Existing Solid Models

112. Left click on the green checkmark as shown in Figure 107.

Figure 107

113. Your screen should look similar to Figure 108

Figure 108

250

114. Move the cursor to the middle left portion of the screen where the part tree is located. Right click on **CirPattern1**. A pop up menu will appear. Left click on the **Suppress** icon as shown in Figure 109.

Figure 109

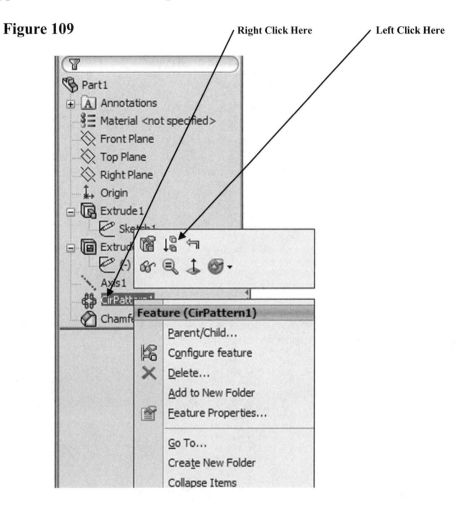

Chapter 5: Learning To Edit Existing Solid Models

115. All holes created using the Circular Pattern command will be suppressed except for the original hole as shown in Figure 110.

Figure 110

116. SolidWorks will gray out the text in the part tree as shown in Figure 111. Use the same steps to "unsuppress" the CirPattern1.

Figure 111

Text Grayed Out

117. The names of all branches in the part tree can also be edited. Left click once on the text **Boss-Extude1/Extrude1**. The text will become highlighted as shown in Figure 112. After the text is highlighted, left click once. The text can now be edited.

Figure 112

118. Enter the text **Base Extrusion** as shown in Figure 113. Press **Enter** on the keyboard. Text for each of the individual operations can be edited if desired.

Figure 113

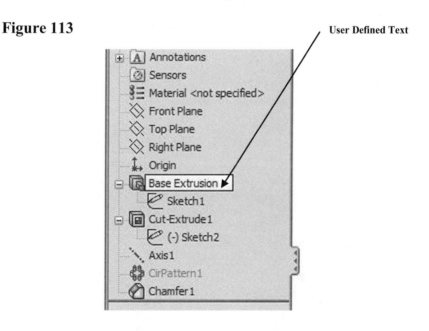

Chapter 5: Learning To Edit Existing Solid Models

119. Notice that the final design looks significantly different than the original design. The new part was redesigned by modifying the existing part as shown in Figure 114.

Figure 114

Drawing Activities

Use these problems from Chapters 1 and 2 to create redesigned parts.

Problem 1

Problem 2

Chapter 5: Learning To Edit Existing Solid Models

Problem 3

Problem 4

Problem 5

Problem 6

Problem 7

Problem 8

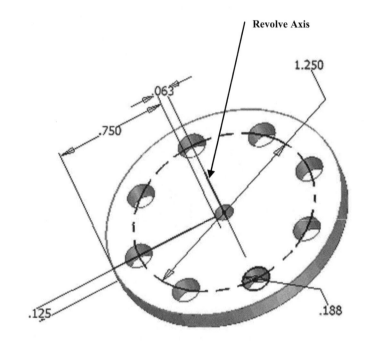

Chapter 6 Advanced Design Procedures

Objectives:
- Design multiple sketch parts
- Learn to use the Front, Top, and Right Planes
- Learn to use the Shell command
- Learn to use the Wireframe viewing command

Chapter 6 includes instruction on how to design the parts shown below.

Chapter 6: Advanced Design Procedures

1. Start SolidWorks by referring to "Chapter 1 Getting Started".

2. After SolidWorks is running, begin a new sketch.

3. Move the cursor to the upper middle portion of the screen and left click on **Circle** as shown in Figure 1.

Figure 1

4. Move the cursor to the center of the screen and left click once. This will be the center of the circle as shown in Figure 2.

Figure 2

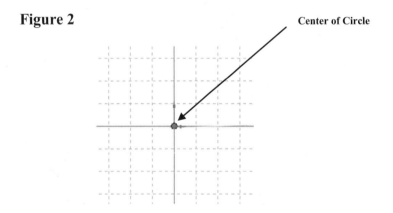

5. Move the cursor to the right and left click once as shown in Figure 3.

Figure 3

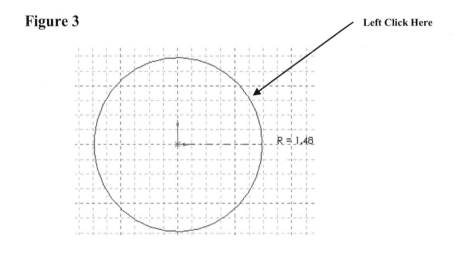

6. Right click anywhere around the drawing. A pop up menu will appear. Left click on **Select** as shown in Figure 4.

Figure 4

7. Move the cursor to the upper left portion of the screen and left click on **Smart Dimension** as shown in Figure 5.

Figure 5

8. After selecting **Smart Dimension** move the cursor over the edge of the circle causing it to turn red as shown in Figure 6. Left click once. The dimension will be attached to the cursor.

Figure 6

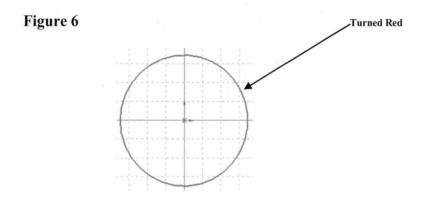

Chapter 6: Advanced Design Procedures

9. Move the cursor to where the dimension will be placed and left click once as shown in Figure 7.

Figure 7

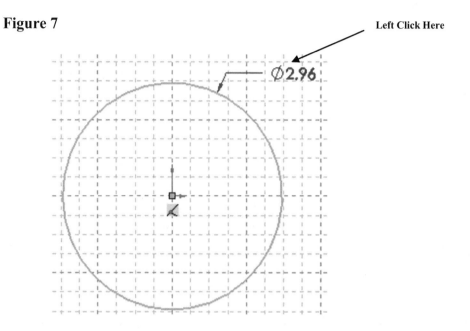

Left Click Here

10. The Modify dialog box will appear as shown in Figure 8.

Figure 8

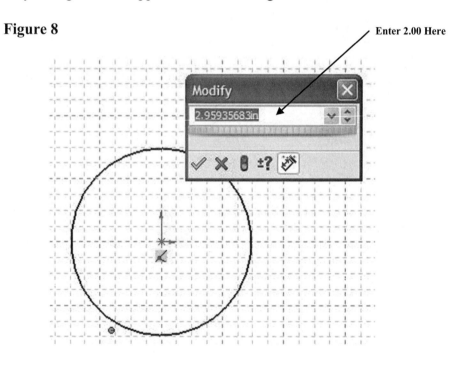

Enter 2.00 Here

Chapter 6: Advanced Design Procedures

11. To edit the dimension, enter **2.00** in the Modify dialog box (while the current dimension is highlighted). Left click on the green checkmark as shown in Figure 9.

Figure 9   Left Click Here

12. The dimension of the circle will become 2.00 inches as shown in Figure 10.

Figure 10

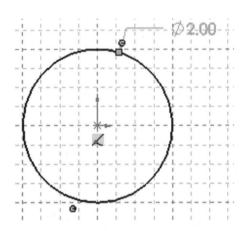

13. Move the cursor to the upper left portion of the screen and left click on the green checkmark as shown in Figure 11.

Figure 11   Left Click Here

263

Chapter 6: Advanced Design Procedures

14. In order to view the entire drawing it may be necessary to move the cursor to the middle portion of the screen and left click once on the "Zoom To Fit" icon as shown in Figure 12.

Figure 12 — Left Click Here

15. The drawing will "fill up" the entire screen. If the drawing is still too large, left click on the "Zoom In/Out" icon as shown in Figure 13. After selecting the Zoom In/Out icon, hold the left mouse button down and drag the cursor up or down to achieve the desired view of the sketch.

Figure 13 — Left Click Here

16. Right click anywhere on the sketch. A pop up menu will appear. Left click on **Exit Sketch** as shown in Figure 14.

Figure 14 — Left Click Here

264

Chapter 6: Advanced Design Procedures

17. SolidWorks is now out of the Sketch commands and into the Features commands. Notice that the commands at the top of the screen are now different. To work in the Features commands a sketch must be present and have no opens (non-connected lines). If there are any opens in the sketch an error message will appear. Your screen should look similar to Figure 15.

Figure 15

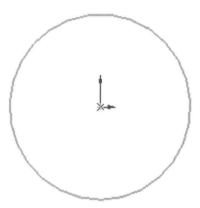

18. Move the cursor to the upper right portion of the screen and left click on the drop down arrow next to the "View Orientation" icon. A drop down menu will appear. Left click on **Trimetric** as shown in Figure 16.

Figure 16   **Left Click Here**

265

Chapter 6: Advanced Design Procedures

19. The view will become trimetric as shown in Figure 17.

Figure 17

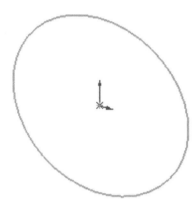

20. Move the cursor to the upper left portion of the screen and left click on **Extruded Boss/Base**. If the Extruded Boss/Base icon did not appear, left click on the Features tab at the upper left portion of the screen. A preview of the extrusion will be displayed.

21. Enter **2.00** for D1 and left click on the green checkmark as shown in Figure 18.

Figure 18

Chapter 6: Advanced Design Procedures

22. Left click anywhere around the part. Your screen should look similar to Figure 19.

Figure 19

23. Move the cursor to the upper left portion of the screen causing the text "Right Plane" to become highlighted as shown in Figure 20.

Figure 20   Text Highlighted

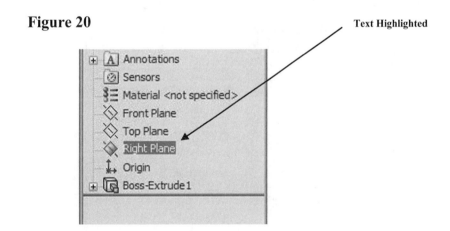

267

Chapter 6: Advanced Design Procedures

24. The Right Plane will become visible as shown in Figure 21.

Figure 21

25. Right click on the **Right Plane** text. A pop up menu will appear. Left click on the **Insert Sketch** icon as shown in Figure 22.

Figure 22

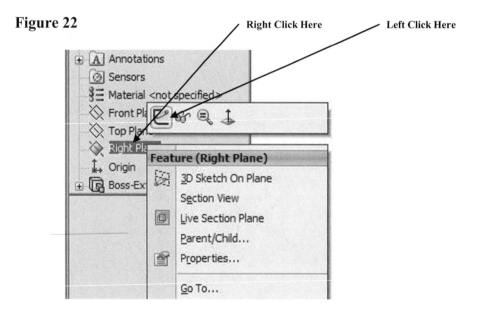

268

26. Your screen should look similar to Figure 23.

   Figure 23

   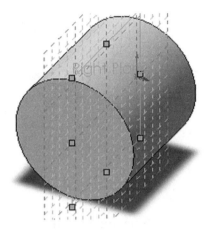

27. Move the cursor to the upper middle portion of the screen and left click on the "Wire Frame Display" icon as shown in Figure 24.

   Figure 24   Left Click Here

28. Your screen should look similar to Figure 25.

   Figure 25

   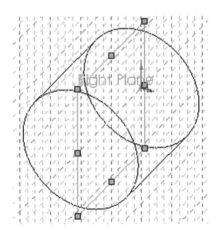

Chapter 6: Advanced Design Procedures

29. Move the cursor to the upper right portion of the screen and left click on the drop down arrow next to the "View Orientation" icon. A drop down menu will appear. Left click on **Right** as shown in Figure 26.

    Figure 26

30. The Right Plane will rotate to a perpendicular view as shown in Figure 27.

    Figure 27

Chapter 6: Advanced Design Procedures

31. The lines that make up the part may not be visible. Move the cursor around where they were located. The lines will appear in red. Move the cursor over the top line causing it to turn red and left click once as shown in Figure 28.

Figure 28

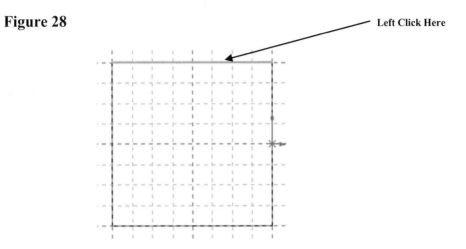

Left Click Here

32. While the line is still green move the cursor to the upper middle portion of the screen and left click on **Convert Entities** as shown in Figure 29. The line will become visible.

Figure 29

Left Click Here

33. Move the cursor over the left vertical line causing it to turn red. Left click once as shown in Figure 30.

Figure 30

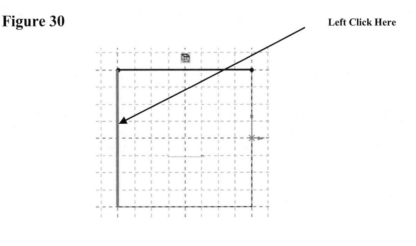

Left Click Here

271

Chapter 6: Advanced Design Procedures

34. While the line is still green move the cursor to the upper middle portion of the screen and left click on **Convert Entities** as shown in Figure 31. The line will become visible on the new plane. You may want to slightly rotate the view isometric to see the line projected on to the new plane.

Figure 31

35. Move the cursor over the lower line causing it to turn red. Left click once as shown in Figure 32.

Figure 32

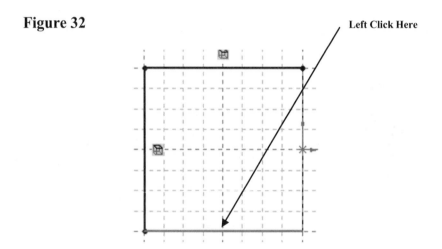

36. While the line is still green move the cursor to the upper middle portion of the screen and left click on **Convert Entities** as shown in Figure 33. The line will become visible.

Figure 33

272

Chapter 6: Advanced Design Procedures

37. Move the cursor over the right vertical line causing it to turn red. Left click once as shown in Figure 34.

Figure 34

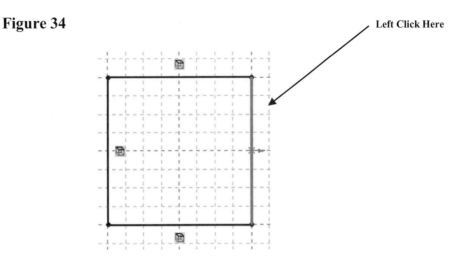

Left Click Here

38. While the line is still green move the cursor to the upper right portion of the screen and left click on **Convert Entities** as shown in Figure 35. The line will become visible.

Figure 35

Left Click Here

39. Your screen should look similar to Figure 36.

Figure 36

273

Chapter 6: Advanced Design Procedures

40. Move the cursor to the upper left portion of the screen and left click on **Line** as shown in Figure 37.

Figure 37

41. Move the cursor to the midpoint of the upper line causing a red dot to appear. Left click as shown in Figure 38.

Figure 38

42. Move the cursor down to the lower line causing a red dot to appear. Left click as shown in Figure 39.

Figure 39

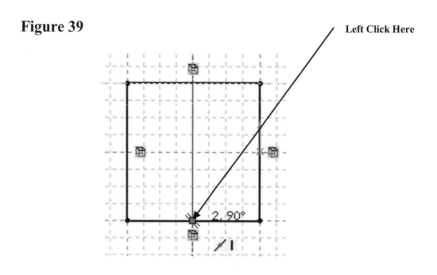

274

Chapter 6: Advanced Design Procedures

43. Right click anywhere around the drawing. A pop up menu will appear. Left click on **Select** as shown in Figure 40.

Figure 40

44. Your screen should look similar to Figure 41.

Figure 41

45. Move the cursor to the upper left portion of the screen and left click on **Circle** as shown in Figure 42.

Figure 42

Chapter 6: Advanced Design Procedures

46. Move the cursor to the midpoint of the center line. A yellow dot will appear. Left click as shown in Figure 43.

Figure 43

47. Move the cursor out to the side and left click as shown in Figure 44.

Figure 44

276

Chapter 6: Advanced Design Procedures

48. Right click anywhere around the drawing. A pop up menu will appear. Left click on **Select** as shown in Figure 45.

Figure 45

49. Move the cursor to the upper left portion of the screen and left click on **Smart Dimension** as shown in Figure 46.

Figure 46

50. Move the cursor over the edge (not center) of the circle until it turns red. Left click once as shown in Figure 47.

Figure 47

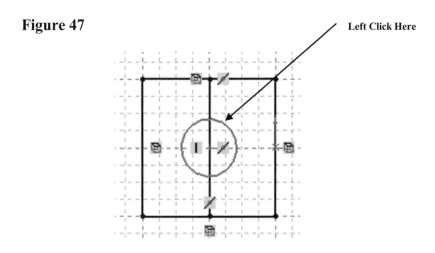

277

Chapter 6: Advanced Design Procedures

51. Move the cursor around. The dimension of the line will appear as shown in Figure 48. The dimension is attached to the cursor.

Figure 48

52. Move the cursor to where the dimension will be placed and left click once. The Modify dialog box will appear as shown in Figure 49.

Figure 49

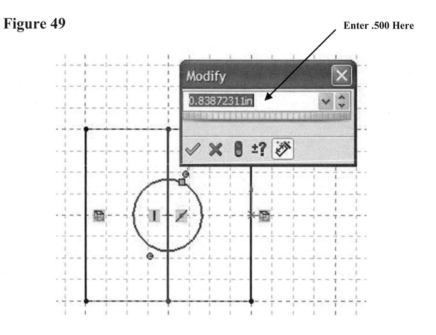

53. To edit the dimension, enter **.500** in the Modify dialog box (while the current dimension is highlighted) and left click on the green checkmark as shown in Figure 50.

Figure 50

54. Your screen should look similar to Figure 51.

Figure 51

55. Right click anywhere around the drawing. A pop up menu will appear. Left click on **Select** as shown in Figure 52.

Figure 52

279

Chapter 6: Advanced Design Procedures

56. You will need to delete each line that was "converted" (5 total) to the new plane. Move the cursor over each line causing it to turn red. Right click on the line. A pop up menu will appear. Left click on **Delete** as shown in Figure 53.

Figure 53

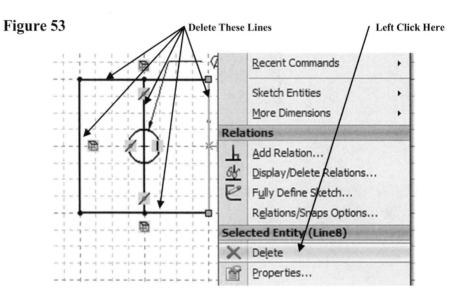

57. Your screen should look similar to Figure 54.

Figure 54

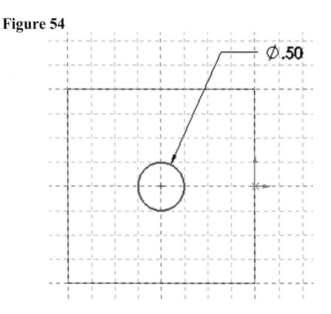

280

58. Move the cursor to the upper right portion of the screen and left click on the "Shaded with Edges" icon as shown in Figure 55.

Figure 55

59. Your screen should look similar to Figure 56.

Figure 56

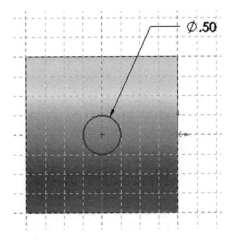

60. After you have verified that no commands are active, right click anywhere on the sketch. A pop up menu will appear. Left click on **Exit Sketch** as shown in Figure 57.

Figure 57

Chapter 6: Advanced Design Procedures

61. Move the cursor to the upper right portion of the screen and left click on the drop down arrow next to the "View Orientation" icon. A drop down menu will appear. Left click on **Trimetric** as shown in Figure 58.

Figure 58

62. Your screen should look similar to Figure 59.

Figure 59

63. Move the cursor to the upper left portion of the screen and left click on **Extruded Cut** as shown in Figure 60. The Extrude dialog box will appear. Move the cursor to the edge of the small circle causing it to turn red. Left click once.

Figure 60

64. Move the cursor to the middle left portion of the screen. Enter **1.00** for D1. Left click in the box next to the text "Direction 2". Enter **1.00** for D2 as shown in Figure 61.

Figure 61

65. A preview will be displayed as shown in Figure 62.

Figure 62

283

Chapter 6: Advanced Design Procedures

66. Move the cursor to the upper left portion of the screen and left click on the green checkmark as shown in Figure 63.

Figure 63

67. Your screen should look similar to Figure 64.

Figure 64

68. Move the cursor to the upper middle portion of the screen and left click on **Shell** as shown in Figure 65.

Figure 65

69. Left click on the lower surface as shown in Figure 66.

Figure 66

70. Enter **.10** next to D1. Left click on the green checkmark as shown in Figure 67.

Figure 67

71. Your screen should look similar to Figure 68.

Figure 68

Chapter 6: Advanced Design Procedures

72. Move the cursor to the upper right portion of the screen and left click on the drop down arrow next to the "View Orientation" icon. A drop down menu will appear. Left click on **Front** as shown in Figure 69.

Figure 69

73. The part will rotate to provide a perpendicular view of the inside as shown in Figure 70.

Figure 70

Chapter 6: Advanced Design Procedures

74. Move the cursor over the same surface causing both the inside and outside lines to turn red. You may have to zoom in for SolidWorks to find both lines at the same time. Both lines must be red at the same time. After both lines are red at the same time, right click on the surface as shown in Figure 71. The surface will turn green.

Figure 71

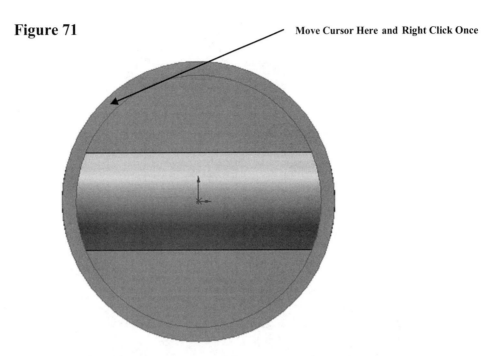

75. A pop up menu will appear. Left click on **Insert Sketch** as shown in Figure 72.

Figure 72

287

Chapter 6: Advanced Design Procedures

76. A new sketch will appear on the selected surface. Your screen should look similar to Figure 73.

Figure 73

77. Move the cursor to the upper left portion of the screen and left click on **Line** as shown in Figure 74.

Figure 74   Left Click Here

78. Move the cursor to the center of the part causing a red dot to appear. Left click as shown in Figure 75.

Figure 75   Left Click Here

288

Chapter 6: Advanced Design Procedures

79. Move the cursor upward and left click as shown in Figure 76.

   Figure 76

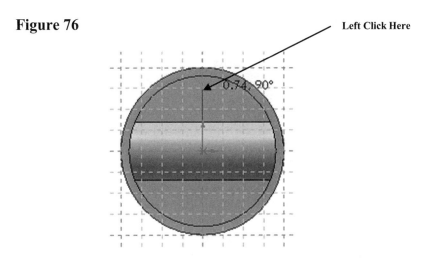

80. Right click anywhere around the drawing. A pop up menu will appear. Left click on **Select** as shown in Figure 77.

   Figure 77

81. Move the cursor to the upper left portion of the screen and left click on **Line** as shown in Figure 78.

   Figure 78

289

Chapter 6: Advanced Design Procedures

82. Move the cursor to the center of the part causing a red dot to appear. Left click as shown in Figure 79.

Figure 79

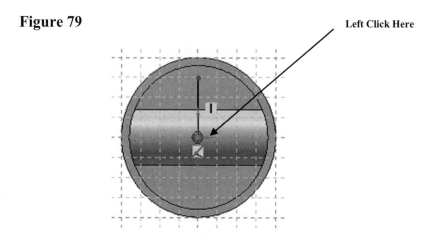

Left Click Here

83. Move the cursor to the left and left click as shown in Figure 80.

Figure 80

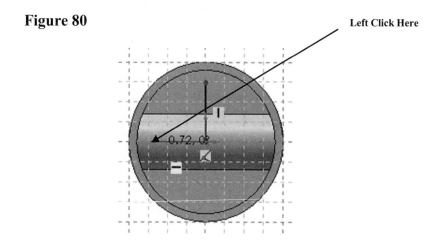

Left Click Here

84. Right click anywhere around the drawing. A pop up menu will appear. Left click on **Select** as shown in Figure 81.

Figure 81

Left Click Here

Chapter 6: Advanced Design Procedures

85. Your screen should look similar to Figure 82.

   Figure 82

86. Move the cursor to the upper left portion of the screen and left click on **Line** as shown in Figure 83.

   Figure 83   Left Click Here

87. Move the cursor to the position shown in Figure 84 and left click once.

   Figure 84   Left Click Here

Chapter 6: Advanced Design Procedures

88. Move the cursor downward and left click as shown in Figure 85.

Figure 85

88. Move the cursor downward and left click as shown in Figure 85.

89. Move the cursor to the right and left click as shown in Figure 86.

Figure 86

90. Move the cursor upward. Ensure that dots appear from the original starting point. Left click as shown in Figure 87.

Figure 87

91. Move the cursor to the left and left click as shown in Figure 88.

Figure 88

92. Right click anywhere around the drawing. A pop up menu will appear. Left click on **Select** as shown in Figure 89.

Figure 89

93. Your screen should look similar to Figure 90.

Figure 90

293

Chapter 6: Advanced Design Procedures

94. Move the cursor to the upper left portion of the screen and left click on **Smart Dimension** as shown in Figure 91.

Figure 91

95. After selecting **Smart Dimension** move the cursor over the lower endpoint of the vertical line coming out of the center of the part. A red dot will appear. Left click on the endpoint as shown in Figure 92.

Figure 92

96. Move the cursor to the endpoint of the far left line. A red dot will appear. Left click once as shown in Figure 93.

Figure 93

97. Move the cursor upward. The dimension of the line will appear as shown in Figure 94. The dimension is attached to the cursor.

Figure 94

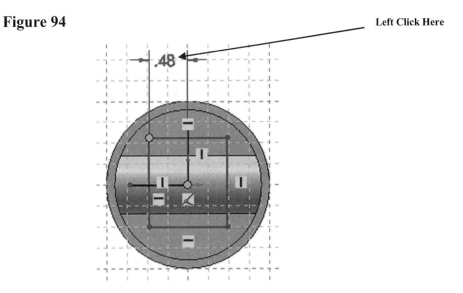

Left Click Here

98. Move the cursor to where the dimension will be placed and left click once. The Modify dialog box will appear as shown in Figure 95.

Figure 95

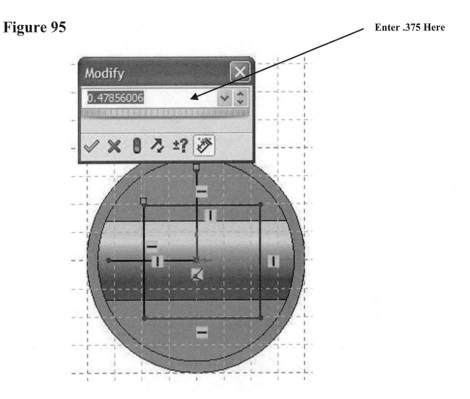

Enter .375 Here

Chapter 6: Advanced Design Procedures

99. To edit the dimension, enter **.375** in the Modify dialog box (while the current dimension is highlighted). Left click on the green checkmark as shown in Figure 96.

Figure 96

100. The dimension of the line will become .375 inches as shown in Figure 97.

Figure 97

101. Move the cursor to the upper left portion of the screen and left click on **Smart Dimension** as shown in Figure 98.

Figure 98

296

Chapter 6: Advanced Design Procedures

102. Move the cursor over the end point of the horizontal line coming out of the center. Left click on the endpoint as shown in Figure 99.

Figure 99

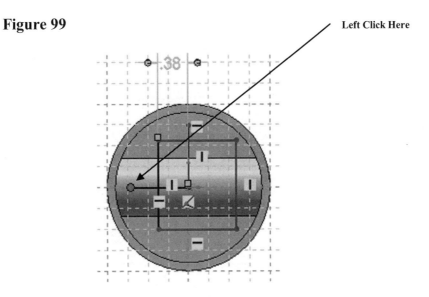

103. Move the cursor to the endpoint of the upper line and left click once as shown in Figure 100.

Figure 100

297

104. Move the cursor out to the side. The dimension of the line will appear as shown in Figure 101. The dimension is attached to the cursor.

Figure 101

105. Move the cursor to where the dimension will be placed and left click once. The Modify dialog box will appear as shown in Figure 102.

Figure 102

Enter .500 Here

Chapter 6: Advanced Design Procedures

106. To edit the dimension, enter **.500** in the Modify dialog box (while the current dimension is highlighted). Left click on the green checkmark as shown in Figure 103.

Figure 103

107. The dimension of the line will become .500 inches as shown in Figure 104.

Figure 104

108. Move the cursor to the upper left portion of the screen and left click on **Smart Dimension** as shown in Figure 105.

Figure 105

299

Chapter 6: Advanced Design Procedures

109. Move the cursor over the endpoint of the horizontal line coming out of the center of the part. Left click on the endpoint as shown in Figure 106.

Figure 106

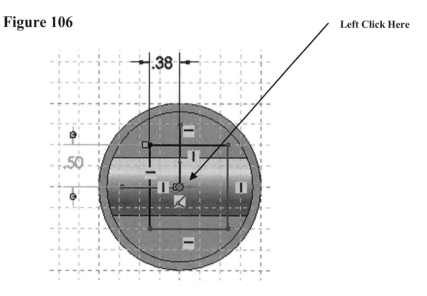

110. Move the cursor to the endpoint of the lower line and left click once as shown in Figure 107.

Figure 107

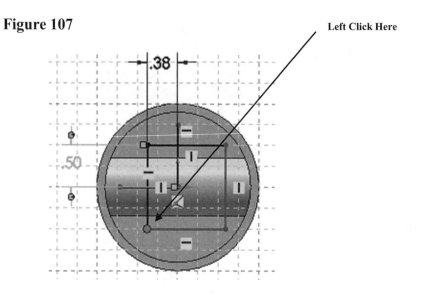

111. Move the cursor out to the side. The dimension of the line will appear as shown in Figure 108. The dimension is attached to the cursor.

Figure 108

112. Move the cursor to where the dimension will be placed and left click once. The Modify dialog box will appear as shown in Figure 109.

Figure 109

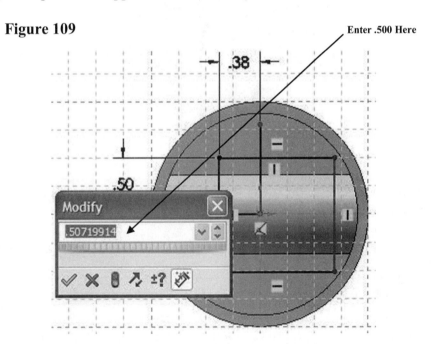

Chapter 6: Advanced Design Procedures

113. To edit the dimension, enter **.500** in the Modify dialog box (while the current dimension is highlighted). Left click on the green checkmark as shown in Figure 110.

Figure 110

114. The dimension of the line will become .500 inches as shown in Figure 111.

Figure 111

115. Move the cursor to the upper left portion of the screen and left click on **Smart Dimension** as shown in Figure 112.

Figure 112

302

Chapter 6: Advanced Design Procedures

116. After selecting **Smart Dimension** move the cursor to the upper endpoint of the vertical line coming out of the center of the part. Left click as shown in Figure 113.

Figure 113

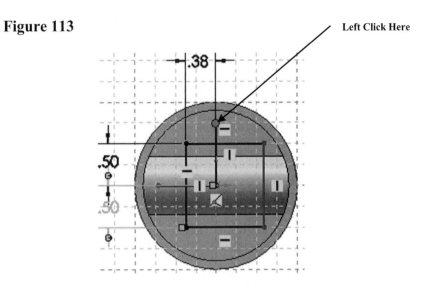

117. Move the cursor to the endpoint of the right vertical line and left click once as shown in Figure 114.

Figure 114

Chapter 6: Advanced Design Procedures

118. Move the cursor out to the side. The dimension of the line will appear as shown in Figure 115. The dimension is attached to the cursor.

Figure 115

119. Move the cursor to where the dimension will be placed and left click once. The Modify dialog box will appear as shown in Figure 116.

Figure 116   Enter .375 Here

304

Chapter 6: Advanced Design Procedures

120. To edit the dimension, enter **.375** in the Modify dialog box (while the current dimension is highlighted) and left click on the green checkmark as shown in Figure 117.

Figure 117   Left Click Here

121. The dimension of the line will become .375 inches as shown in Figure 118.

Figure 118

305

Chapter 6: Advanced Design Procedures

122. Right click anywhere around the drawing. A pop up menu will appear. Left click on **Select** as shown in Figure 119.

Figure 119

123. Move the cursor over the vertical line coming out of the center of the part causing it to turn red as shown in Figure 120.

Figure 120

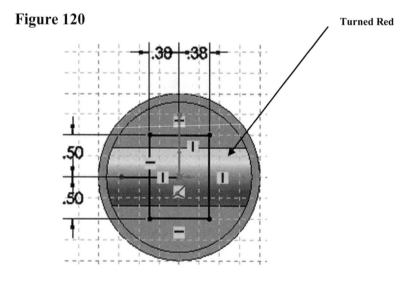

306

Chapter 6: Advanced Design Procedures

124. After the line has turned red, right click once. A pop up menu will appear. Left click on **Delete** as shown in Figure 121.

Figure 121

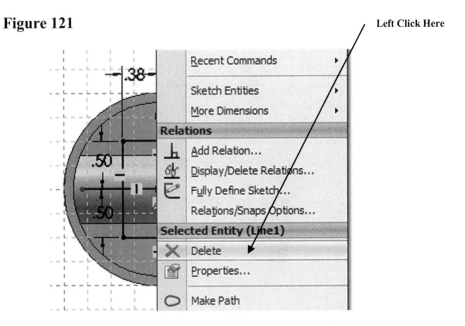

Left Click Here

125. Use the same steps to delete the horizontal line coming out of the center of the part as shown in Figure 122. If the dimensions disappear do not be concerned.

Figure 122

Delete This Line

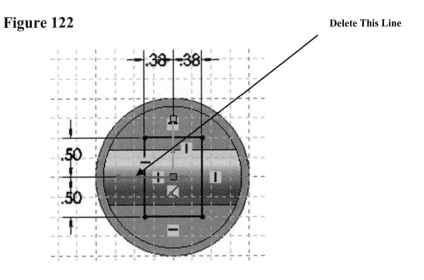

307

Chapter 6: Advanced Design Procedures

126. Right click anywhere around the drawing. A pop up menu will appear. Left click on **Exit Sketch** as shown in Figure 123.

Figure 123

127. Move the cursor to the upper middle portion of the screen and left click on the drop down arrow next to the "View Orientation" icon. A drop down menu will appear. Left click on **Trimetric** as shown in Figure 124.

Figure 124

308

Chapter 6: Advanced Design Procedures

128. Your screen should look similar to Figure 125.

Figure 125

129. Move the cursor to the upper left portion of the screen and left click on **Extruded Cut** as shown in Figure 126. If SolidWorks does not provide a preview of the extruded cut, left click on any of the lines that form the box.

Figure 126

130. Enter **1.875** for D1 as shown in Figure 127.

Figure 127

309

Chapter 6: Advanced Design Procedures

131. Left click on the green checkmark as shown in Figure 128.

Figure 128

132. Your screen should look similar to Figure 129. Use the Rotate command to roll the part around to view the inside.

Figure 129

133. Save the part as Piston1.SLDPRT where it can be easily retired later.

134. Begin a new drawing as described in Chapter 1.

135. Draw a circle in the center of the grid as shown in Figure 130.

Figure 130

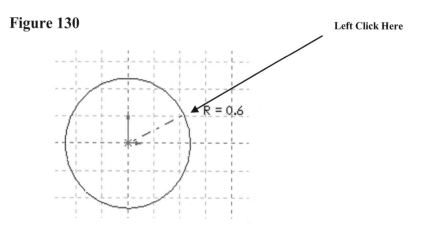

310

Chapter 6: Advanced Design Procedures

136. Use the **Smart Dimension** command to dimension the circle to **.500** inches as shown in Figure 131.

Figure 131

12.7 mm

137. Exit the Sketch commands and Extrude the circle to a length of **1.875** inches as shown in Figure 132.

Figure 132

47.62 mm

311

Chapter 6: Advanced Design Procedures

138. Your screen should look similar to Figure 133.

Figure 133

139. Save the part as Wristpin1.SLDPRT where it can be easily retrieved later.

140. Begin a new sketch as described in Chapter 1.

141. Complete the sketch shown in Figure 134.

Figure 134

142. Exit the Sketch commands and change the view to Trimetric. A trimetric view will be displayed as shown in Figure 135.

Figure 135

143. Extrude the sketch to a distance of **2.25** inches. Your screen should look similar to Figure 136. 57.15mm

Figure 136

Chapter 6: Advanced Design Procedures

144. Use the **Fillet** command to create **1.125** inch fillets on the front portion of the part as shown in Figure 137.

Figure 137

145. Your screen should look similar to Figure 138.

Figure 138

146. Move the cursor to the surface shown in Figure 139 causing it to turn red.

Figure 139

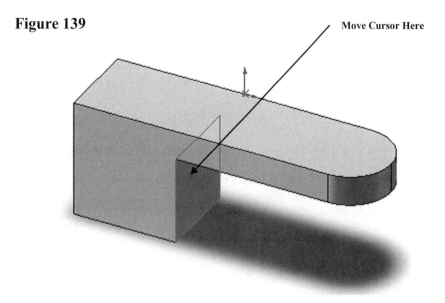

147. Left click once as shown in Figure 140. The surface will turn green.

Figure 140

Chapter 6: Advanced Design Procedures

148. Right click on the surface. A pop up menu will appear. Left click on **Insert Sketch** as shown in Figure 141.

Figure 141

149. Move the cursor to the upper middle portion of the screen and left click on the drop down arrow next to the "Views Orientation" icon. A drop down menu will appear. Left click on **Normal To** as shown in Figure 142.

Figure 142

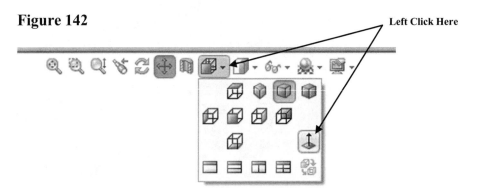

Chapter 6: Advanced Design Procedures

150. A perpendicular view of the surface will be displayed as shown in Figure 143.

Figure 143

151. Create a sketch on the selected surface as shown in Figure 144.

Figure 144

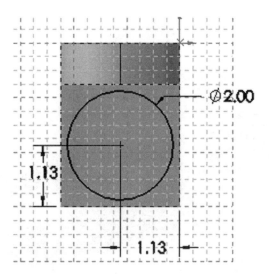

Chapter 6: Advanced Design Procedures

152. Exit out of the Sketch commands and change the view to trimetric as shown in Figure 145.

Figure 145

153. Use the **Extruded Cut** command to extrude or "cut" out (clear through the part) the circle that was just completed. Your screen should look similar to Figure 146.

Figure 146

318

Chapter 6: Advanced Design Procedures

154. Left click on the top portion of the part causing it to turn green as shown in Figure 147.

Figure 147

Left Click Here

155. Move the cursor to the upper middle portion of the screen and left click on the drop down arrow next to the "View Orientation" icon. A drop down menu will appear. Left click on **Normal To** as shown in Figure 148.

Figure 148

Left Click Here

Chapter 6: Advanced Design Procedures

156. A perpendicular view will be displayed as shown in Figure 149.

Figure 149

157. Draw the sketch as shown in Figure 150.

Figure 150

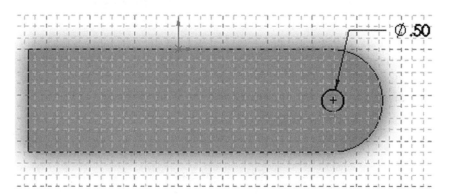

158. Use the **Extruded Cut** command to extrude or "cut" out the circle that was just completed. Your screen should look similar to Figure 151.

Figure 151

159. Change the view to trimetric. Your screen should look similar to Figure 152.

Figure 152

Part 5

160. Save the part as Pistoncase1.SLDPRT where it can be easily retrieved later.

161. Begin a new drawing as described in Chapter 1.

162. Complete the sketch shown in Figure 153.

Figure 153

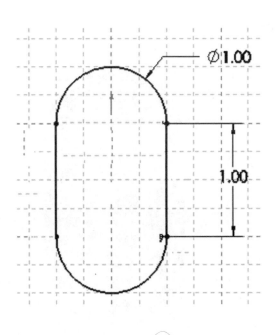

321

Chapter 6: Advanced Design Procedures

163. Extrude the sketch into a solid with a thickness of **.25** as shown in Figure 154.

Figure 154

164. Complete the following sketch. Use the center of the outside fillet radius as the center of the circle as shown in Figure 155.

Figure 155

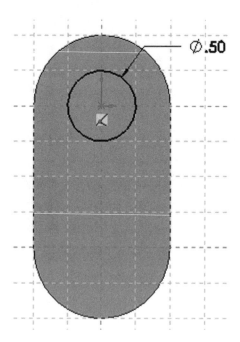

165. Extrude the sketch into a solid with a thickness of **.25** as shown in Figure 156.

Figure 156

166. Use the rotate command and roll the part around to gain access to the opposite side as shown in Figure 157.

Figure 157

Chapter 6: Advanced Design Procedures

167. Begin a new sketch on the opposite side as shown in Figure 158.

Figure 158

168. Use the **Normal To** command to gain a perpendicular view as shown in Figure 159.

Figure 159

169. Complete the following sketch as shown in Figure 160.

Figure 160

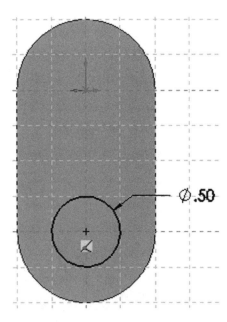

170. Extrude the sketch into a solid with a thickness of **.25** as shown in Figure 161.

Figure 161

171. Save the part as Crankshaft1.SLDPRT where it can be easily retrieved later.

172. Begin a new drawing as described in Chapter 1.

Chapter 6: Advanced Design Procedures

173. Complete the sketch shown in Figure 162. Extrude the sketch into a solid with a thickness of **.25**.

Figure 162

174. Your screen should look similar to Figure 163.

Figure 163

175. Save the part as Conrod1.SLDPRT where it can be easily retrieved later.

176. All of these parts will be used in the next chapter.

Chapter 7 Introduction to Assembly View Procedures

Objectives:
- Learn to import existing solid models into an Assembly
- Learn to constrain all parts in an Assembly
- Learn to edit/modify parts while in an Assembly
- Learn to use the Motion Study command in an Assembly

Chapter 7 includes instruction on how to construct the assembly shown below.

Chapter 7: Introduction to Assembly View Procedures

1. Start SolidWorks by referring to "Chapter 1 Getting Started".

2. After SolidWorks is running, begin an assembly drawing. Move the cursor to the upper left corner of the screen and left on **New**. The New SolidWorks Document dialog box will appear. Left click on **Assembly** as shown in Figure 1.

Figure 1

3. Left click on **OK**.

Chapter 7: Introduction to Assembly View Procedures

4. The Assembly feature of SolidWorks will open. Left click on the **Assembly** tab as shown. Your screen should look similar to Figure 2.

Figure 2

5. Move the cursor to the lower left portion of the screen and left click **Browse** as shown in Figure 3.

Figure 3

Chapter 7: Introduction to Assembly View Procedures

6. The Open dialog box will appear. Locate the Case1.SLDPART file and left click on **Open** as shown in Figure 4.

Figure 4

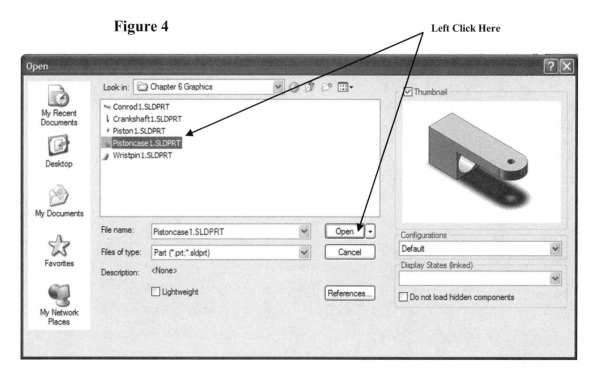

7. SolidWorks will place one piston case in the drawing space. The piston case will be attached to the cursor. The piston case will appear as shown in Figure 5. If the part does not appear attached to the cursor simply left click anywhere in the work area.

Figure 5

330

8. Left click once. The piston case will be placed in the assembly drawing. Your screen should look similar to Figure 6.

Figure 6

9. Move the cursor to the upper left portion of the screen and left click on **Insert Components**. If the Assembly tools are not visible, left click on the **Assembly** tab as shown in Figure 7.

Figure 7   Left Click Here

10. Move the cursor to the lower left portion of the screen and left click on **Browse** as shown in Figure 8.

Figure 8   Left Click Here

Chapter 7: Introduction to Assembly View Procedures

11. The Open dialog box will appear. Locate the Piston1.SLDPRT file and left click on **Open** as shown in Figure 9.

Figure 9

12. The piston will be attached to the cursor. Place the piston anywhere near the piston case and left click once. Your screen should look similar to Figure 10.

Figure 10

332

Chapter 7: Introduction to Assembly View Procedures

13. Move the cursor to the upper left portion of the screen and left click on **Insert Components** as shown in Figure 11.

Figure 11

14. Move the cursor to the lower left portion of the screen and left click on **Browse** as shown in Figure 12.

Figure 12

333

Chapter 7: Introduction to Assembly View Procedures

15. The Open dialog box will appear. Locate the Conrod1.SLDPRT file and left click on **Open** as shown in Figure 13.

Figure 13

16. The connecting rod will be attached to the cursor. Place the connecting rod anywhere near the piston case and left click once. On the keyboard press the **Esc** button on the key board once. Your screen should look similar to Figure 14.

Figure 14

Chapter 7: Introduction to Assembly View Procedures

17. Move the cursor to the upper left portion of the screen and left click on **Insert Components** as shown in Figure 15.

Figure 15

18. Move the cursor to the lower left portion of the screen and left click on **Browse** as shown in Figure 16.

Figure 16

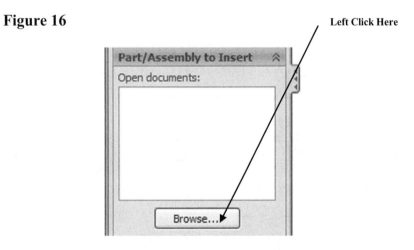

Chapter 7: Introduction to Assembly View Procedures

19. The Open dialog box will appear. Locate the crankshaft1.SLDPRT file and left click on **Open** as shown in Figure 17.

Figure 17

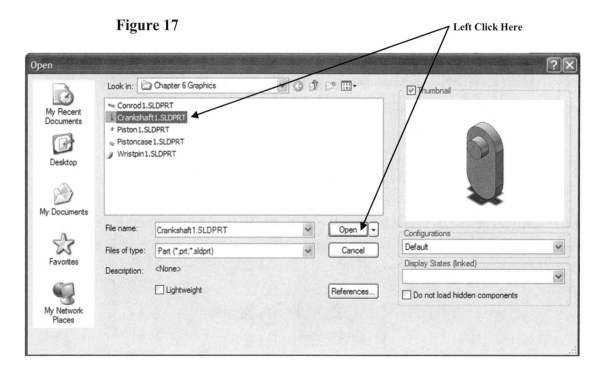

20. The crankshaft will be attached to the cursor. Place the crankshaft anywhere near the piston case and left click once. Your screen should look similar to Figure 18.

Figure 18

Chapter 7: Introduction to Assembly View Procedures

21. Move the cursor to the upper left portion of the screen and left click on **Insert Components** as shown in Figure 19.

Figure 19

22. Move the cursor to the lower left portion of the screen and left click on **Browse** as shown in Figure 20.

Figure 20

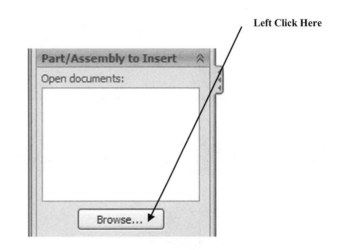

337

Chapter 7: Introduction to Assembly View Procedures

23. The Open dialog box will appear. Locate the Wristpin1.SLDPRT file and left click on **Open** as shown in Figure 21.

Figure 21

24. The wristpin will be attached to the cursor. Place the wristpin anywhere near the piston case and left click once. Your screen should look similar to Figure 22.

Figure 22

338

Chapter 7: Introduction to Assembly View Procedures

25. The first part inserted into the assembly becomes "fixed" meaning it cannot be moved. All other parts can be moved by left clicking (holding the left mouse button down) and dragging the part to the desired location.

26. Use the **Rotate** command (or press the mouse wheel down) to rotate the entire parts group to the position shown in Figure 23.

Figure 23

27. Move the cursor to the upper middle portion of the screen and left click on **Rotate Component** as shown in Figure 24.

Figure 24

339

Chapter 7: Introduction to Assembly View Procedures

28. Left click (holding the left mouse button down) and rotate the piston to a horizontal position as shown in Figure 25.

Figure 25 Horizontal Position

29. Right click anywhere around the part. A pop up menu will appear. Left click on **Select** as shown in Figure 26.

Figure 26 Left Click Here

30. Move the cursor to the upper left portion of the screen and left click on **Mate** as shown in Figure 27.

Figure 27 Left Click Here

340

Chapter 7: Introduction to Assembly View Procedures

31. Move the cursor to the piston causing the OUTSIDE edges of the part to turn red. Left click once. The part will turn green as shown in Figure 28.

Figure 28

Turned Green

32. Move the cursor to the piston case causing the cylinder edges to turn red. Left click once. The cylinder will turn green as shown in Figure 29.

Figure 29

Turned Green

33. The centers of the piston and the cylinder will be aligned. Your screen should look similar to Figure 30.

Figure 30

34. Left click on the green checkmark beneath the text "Concentric1" as shown in Figure 31.

Figure 31 Left Click Here

35. Left click on the green checkmark beneath the text "Mate" as shown in Figure 32.

Figure 32 Left Click Here

Chapter 7: Introduction to Assembly View Procedures

36. Left click (holding the left mouse button down) and drag the piston down into the cylinder as shown in Figure 33.

Figure 33

37. Move the cursor to the lower left portion of the piston. Left click (holding the left mouse button down) and slide the piston downward, out below the bore as shown in Figure 34.

Figure 34

38. Move the cursor to the upper left portion of the screen and left click on **Mate** as shown in Figure 35.

Figure 35

39. Move the cursor to the wristpin hole on the piston causing the inside edge of the hole to turn red. Left click once as shown in Figure 36.

Figure 36

40. Move the cursor to the upper portion of the connecting rod causing the inside edges of the hole to turn red. You may have to drag the connecting rod to a position where the end holes are accessible. Left click as shown in Figure 37.

Figure 37

Left Click Here

Chapter 7: Introduction to Assembly View Procedures

41. The connecting rod hole and the center of the wristpin hole in the piston will be aligned as shown in Figure 38.

Figure 38

42. Left click on the green checkmark beneath the text "Concentric2" as shown in Figure 39.

Figure 39

Left Click Here

43. Left click on the green checkmark beneath the text "Mate" as shown in Figure 40.

Figure 40

Left Click Here

Chapter 7: Introduction to Assembly View Procedures

44. Use the rotate command to rotate the entire assembly around to gain access to the underside of the piston. Move the connecting rod to the location shown in Figure 41. This will take some skill. If the piston moves during this process, use the cursor to move the piston to the original location.

Figure 41

45. Move the cursor to the upper left portion of the screen and left click on **Mate** as shown in Figure 42.

Figure 42   Left Click Here

46. Move the cursor to the left side of the connecting rod causing the edges of the connecting rod to turn red. Left click as shown in Figure 43. You may have to zoom in so that SolidWorks will find the proper surface.

Figure 43   Left Click Here

Chapter 7: Introduction to Assembly View Procedures

47. Press the mouse wheel down to rotate the assembly around to gain access to the surface opposite the previously selected surface as shown in Figure 44. Select the second mating surface (while both edges are red) by left clicking as shown in Figure 44.

Figure 44

48. Left click on the Distance icon and enter **.25** as shown in Figure 45.

Figure 45

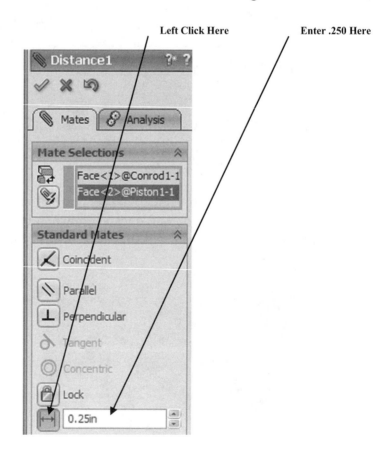

348

Chapter 7: Introduction to Assembly View Procedures

49. Left click on the green checkmark beneath the text "Distance1" as shown in Figure 46.

Figure 46

50. Left click on the green checkmark beneath the text "Mate" as shown in Figure 47.

Figure 47

51. The connecting rod should be centered in the piston. Your screen should look similar to Figure 48.

Figure 48

Chapter 7: Introduction to Assembly View Procedures

52. Rotate the entire assembly to the position shown in Figure 49.

Figure 49

53. Move the cursor to the upper left portion of the screen and left click on **Mate** as shown in Figure 50.

Figure 50

Left Click Here

Chapter 7: Introduction to Assembly View Procedures

54. Move the cursor to the wristpin causing the edges of the wristpin to turn red. Left click once as shown in Figure 51.

Figure 51

55. Move the cursor to the wristpin hole causing the edges of the hole to turn red. Left click once as shown in Figure 52.

Figure 52

56. Left click on the green checkmark beneath the text "Concentric3" as shown in Figure 53.

Figure 53

57. Left click on the green checkmark beneath the text "Mate" as shown in Figure 54.

Figure 54

Chapter 7: Introduction to Assembly View Procedures

58. The wristpin will be placed in line with the wristpin hole as shown in Figure 55.

Figure 55

59. Move the cursor to the upper left portion of the screen and left click on **Mate** as shown in Figure 56.

Figure 56

Left Click Here

60. Move the cursor to the side of the wristpin causing the edges of the wristpin to turn red. Left click once as shown in Figure 57.

Figure 57

Left Click Here

Chapter 7: Introduction to Assembly View Procedures

61. Move the cursor to the side of the connecting rod causing the edges of the connecting rod to turn red. Left click once as shown in Figure 58.

Figure 58

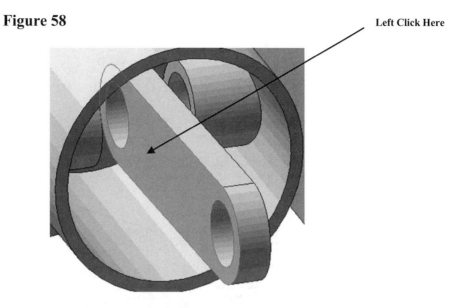

Left Click Here

62. Left click on the Distance icon and enter **.7825**. Left click next to "Flip dimension" as shown in Figure 59.

Figure 59

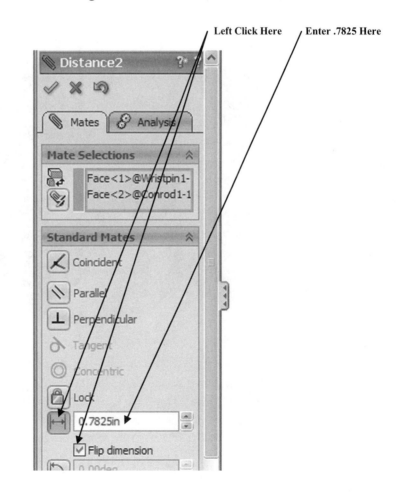

Left Click Here   Enter .7825 Here

353

Chapter 7: Introduction to Assembly View Procedures

63. Left click on the green checkmark beneath the text "Distance2" as shown in Figure 60.

Figure 60

64. Left click on the green checkmark beneath the text "Mate" as shown in Figure 61.

Figure 61

65. Your screen should look similar to Figure 62.

Figure 62

Chapter 7: Introduction to Assembly View Procedures

66. Move the cursor to the upper middle portion of the screen and left click on **Move Component** as shown in Figure 63.

Figure 63

67. Left click again (holding the left mouse button down) and drag the crankshaft to where it can be rotated as shown in Figure 64.

Figure 64

355

Chapter 7: Introduction to Assembly View Procedures

68. Right click anywhere around the parts. A pop up menu will appear. Left click on **Select** as shown in Figure 65.

Figure 65

69. Move the cursor to the upper middle portion of the screen and left click on **Rotate Component** as shown in Figure 66.

Figure 66

Chapter 7: Introduction to Assembly View Procedures

70. Left click once. Left click again (holding the left mouse button down) and rotate the connecting rod upward as shown in Figure 67.

Figure 67

71. Right click anywhere around the parts. A pop up menu will appear. Left click on **Select** as shown in Figure 68.

Figure 68

Chapter 7: Introduction to Assembly View Procedures

72. Move the cursor to the upper left portion of the screen and left click on **Mate** as shown in Figure 69.

Figure 69

73. Move the cursor to the crankshaft pin that will secure the connecting rod. The edges of the crankshaft pin will turn red. Left click once as shown in Figure 70.

Figure 70

74. Move the cursor to the connecting rod hole that will be secured to the crankshaft. The edges of the connecting rod hole will turn red. Left click once as shown in Figure 71.

Figure 71

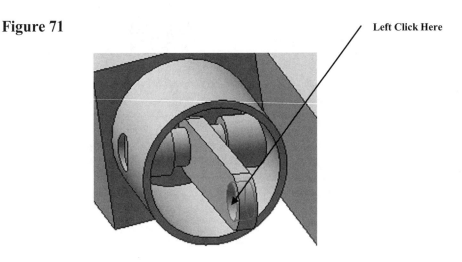

Chapter 7: Introduction to Assembly View Procedures

75. Left click on the green checkmark beneath the text "Concentric4" as shown in Figure 72.

Figure 72

76. Left click on the green checkmark beneath the text "Mate" as shown in Figure 73.

Figure 73

77. The center of the crankshaft and the center of the connecting rod are now aligned.

78. Left click on the crankshaft (holding the left mouse button down) and drag it to the connecting rod as shown in Figure 74.

Figure 74

Chapter 7: Introduction to Assembly View Procedures

79. Move the cursor to the upper left portion of the screen and left click on **Mate** as shown in Figure 75.

Figure 75

Left Click Here

80. Move the cursor to the crankshaft pin causing the edge of the pin to turn red. Left click as shown in Figure 76.

Figure 76

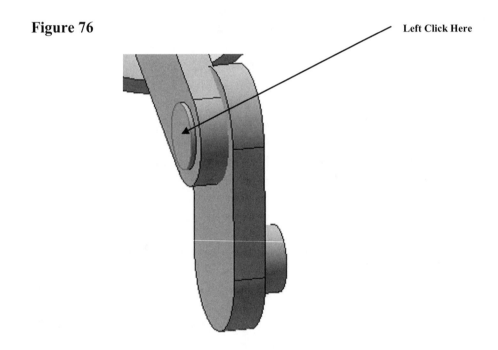

Left Click Here

360

Chapter 7: Introduction to Assembly View Procedures

81. Move the cursor to the connecting rod side. The edge of the connecting rod will turn red. Left click as shown in Figure 77.

Figure 77 — Left Click Here

82. Left click on the green checkmark beneath the text "Coincident3" as shown in Figure 78.

Figure 78 — Left Click Here

361

Chapter 7: Introduction to Assembly View Procedures

83. Left click on the green checkmark beneath the text "Mate" as shown in Figure 79.

Figure 79

84. SolidWorks will place the connecting rod and crankshaft together as shown in Figure 80.

Figure 80

Chapter 7: Introduction to Assembly View Procedures

85. Use the rotate command and roll the assembly around to gain access to the opposite side as shown in Figure 81.

Figure 81

86. Move the cursor to the upper left portion of the screen and left click on **Mate** as shown in Figure 82.

Figure 82

Left Click Here

363

Chapter 7: Introduction to Assembly View Procedures

87. Move the cursor to the crankshaft pin that will be secured in the piston case. The edges of the pin will turn red. Left click once as shown in Figure 83.

Figure 83

88. Move the cursor to the piston case hole that will secure the crankshaft. The edges of the hole will turn red. Left click once as shown in Figure 84.

Figure 84

Chapter 7: Introduction to Assembly View Procedures

89. SolidWorks will place the crankshaft pin into the piston case as shown in Figure 85.

   Figure 85

   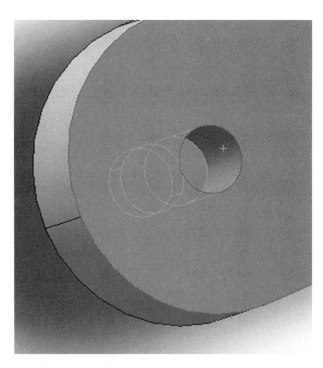

90. Left click on the green checkmark beneath the text "Concentric5" as shown in Figure 86.

   Figure 86

91. Left click on the green checkmark beneath the text "Mate" as shown in Figure 87.

   Figure 87

Chapter 7: Introduction to Assembly View Procedures

92. Your screen should look similar to Figure 88.

Figure 88

93. Rotate the parts around to view the assembly. Your screen should look similar to Figure 89. Save the file at this time.

Figure 89

94. The length of the connecting rod must be modified. Move the cursor over the connecting rod as shown in Figure 90. The edges will turn red.

Figure 90

95. Left click once. The surface of the connecting rod will turn green. Right click once. A pop up menu will appear. Left click on **Edit Sketch** as shown in Figure 91.

Figure 91

96. The sketch used to create the connecting rod will be displayed as shown in Figure 92. Use the Rotate command to obtain a better view of the connecting rod.

Figure 92

97. Your screen should look similar to Figure 93.

Figure 93

98. Move the cursor over the 2.25 dimension. The dimension will turn red. Double click. The Modify dialog box will appear as shown in Figure 94.

Figure 94

Actual Dimension

99. While the text is still highlighted, enter **4.5** and left click on the green checkmark as shown in Figure 95.

Figure 95

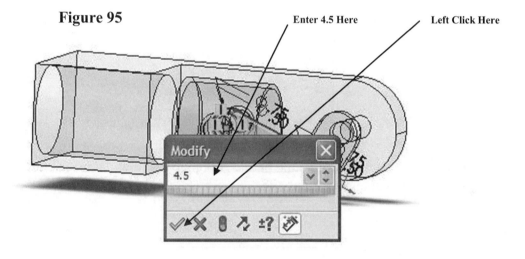

Enter 4.5 Here

Left Click Here

Chapter 7: Introduction to Assembly View Procedures

100. The length of the connecting rod will become 4.50 inches as shown in Figure 96.

Figure 96

New Dimension

101. Left click on the green checkmark beneath the text "Dimension" as shown as shown in Figure 97.

Figure 97

Left Click Here

102. Right click anywhere around the parts. A pop up menu will appear. Left click on **Exit Sketch** as shown in Figure 98.

Figure 98

Left Click Here

Chapter 7: Introduction to Assembly View Procedures

103. Right click anywhere around the parts. A pop up menu will appear. Left click on **Edit Assembly: Assem1** as shown in Figure 99.

Figure 99

104. The changes made to the connecting rod will be displayed. Your screen should look similar to Figure 100.

Figure 100

Chapter 7: Introduction to Assembly View Procedures

105. The length of the crankshaft pin also must be modified. Rotate the assembly to gain access to the backside of the crankshaft as shown in Figure 101.

Figure 101

106. Move the cursor over the crankshaft pin as shown in Figure 101. Ensure that the entire crankshaft pin is highlighted. Left click on the crankshaft pin. The entire pin of the crankshaft will turn green. Right click once. A pop up menu will appear. Left click on **Edit Feature** as shown in Figure 102.

Figure 102

372

107. The existing extrusion length will be displayed as shown in Figure 103.

Figure 103

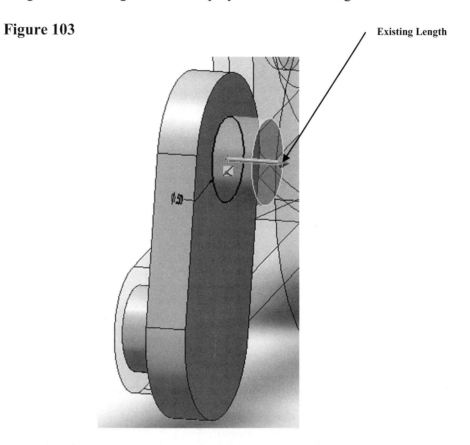

Existing Length

108. Enter **2.00** for the extrusion distance. A preview of the extrusion will be displayed. Left click on the green checkmark as shown in Figure 104.

Figure 104

Enter 2.00 Here — Left Click Here

Chapter 7: Introduction to Assembly View Procedures

109. Right click anywhere around the parts. A pop up menu will appear. Left click on **Edit Assembly: Assem1** as shown in Figure 105.

Figure 105

Left Click Here

110. The changes made to the crankshaft will be displayed. Your screen should look similar to Figure 106.

Figure 106

111. Rotate the part around as shown in Figure 107.

Figure 107

112. Move the cursor to the lower left portion of the screen and left click on **Motion Study 1** as shown in Figure 108.

Figure 108

Chapter 7: Introduction to Assembly View Procedures

113. The Study Motion dialog box will appear. Your screen should look similar to Figure 109.

Figure 109

Chapter 7: Introduction to Assembly View Procedures

114. Move the cursor to the middle right portion of the screen and left click on "Motor" icon as shown in Figure 110.

   Figure 110   Left Click Here

115. Move the cursor to the crankshaft pin that protrudes through the piston case. The crankshaft pin will turn red as shown in Figure 111.

   Figure 111   Left Click Here

377

Chapter 7: Introduction to Assembly View Procedures

116. Left click once. A red arrow will be displayed showing the direction of motion as shown in Figure 112.

Figure 112

117. Left click on the green checkmark beneath the text "Motor" as shown in Figure 113.

Figure 113

118. Move the cursor to the lower left portion of the screen and left click on the "Calculate" icon as shown in Figure 114.

Figure 114

Chapter 7: Introduction to Assembly View Procedures

119. SolidWorks will begin animating the part. If SolidWorks detects any part collision or interference, the animation will stop playing.

120. Move the cursor to the lower left portion of the screen and left click on the "Play" icon as shown in Figure 115.

Figure 115

121. The Animation can be controlled by increasing or decreasing the speed or looping. Left click on the Motion Study Properties icon as shown in Figure 116. The Motion Study Properties panel will appear at the left (not shown). Increase the frames per second to 30 to smooth out the animation. Left click on the minimize icon to enlarge the animation to full screen.

Figure 116

122. Save the assembly where it can easily be retrieved. When closing SolidWorks, a dialog box will appear indicating that models making up the assembly have been modified. Left click on **Yes** as shown in Figure 117.

Figure 117

379

Chapter 8 Introduction to Advanced Commands

Objectives:
- Learn to use the Swept Boss/Base command
- Learn to use the Lofted/Boss Base command
- Learn to use the Plane command

Chapter 8 includes instruction on how to design the parts shown below.

Chapter 8: Introduction to Advanced Commands

1. Start SolidWorks by referring to "Chapter 1 Getting Started".

2. After SolidWorks is running, begin a New Sketch.

3. Move the cursor to the upper left portion of the screen and left click on **Line** as shown in Figure 1.

Figure 1   Left Click Here

4. Move the cursor to the center of the screen and left click once. Ensure that the red dot appears on the intersection of the origin as shown in Figure 2.

Figure 2   Left Click Here

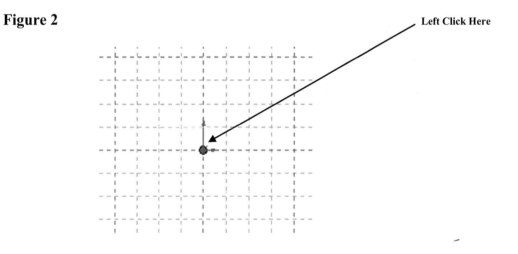

5. Move the cursor to the left portion of the screen and left click once as shown in Figure 3. Right click anywhere on the screen. A pop up menu will appear. Left click on **Select**.

Figure 3

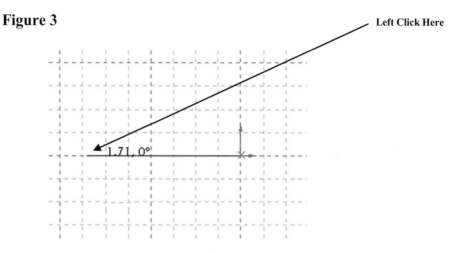

6. Move the cursor to the upper left portion of the screen and left click on **Line** as shown in Figure 4.

Figure 4

Chapter 8: Introduction to Advanced Commands

7. Move the cursor to the center of the grid and left click once as shown in Figure 5. Move the cursor to the left and left click (draw a short line going off to the left). Move the cursor to the upper left portion of the screen and left click on **Rectangle**. Complete the sketch shown in Figure 5. Be sure to start the rectangle at the origin. After the sketch has been dimensioned, delete the horizontal line that was drawn first. Right click anywhere on the screen. A pop up menu will appear. Left click on **Select**.

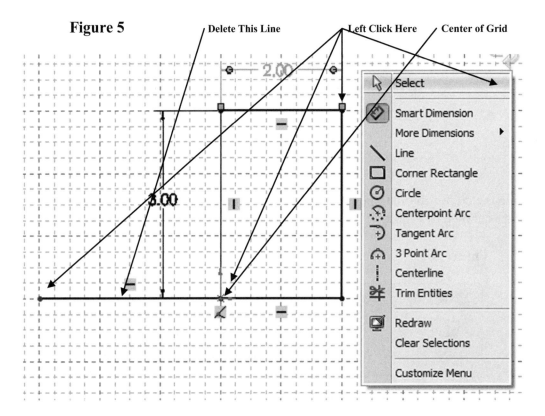

Figure 5

Chapter 8: Introduction to Advanced Commands

8. Right click anywhere on the screen. A pop up menu will appear. Left click **Exit Sketch** as shown in Figure 6.

Figure 6

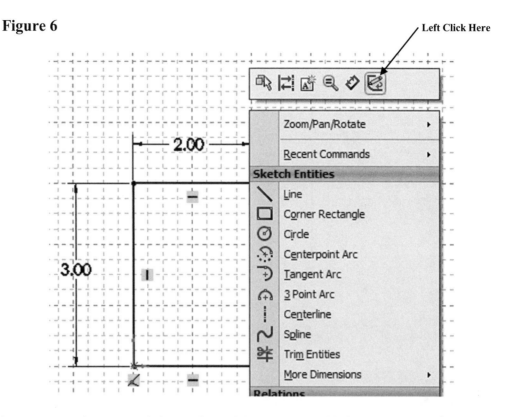

9. Move the cursor to the upper right portion of the screen and left click on the drop down arrow next to the Standard Views icon. A drop down menu will appear. Left click on **Trimetric** as shown in Figure 7.

Figure 7

385

Chapter 8: Introduction to Advanced Commands

10. Your screen should look similar to Figure 8.

   Figure 8

386

11. Move the cursor over the text "Right Plane" causing the text to become highlighted. Notice the right plane is highlighted. Right click once. A pop up menu will appear. Left click on **Sketch** as shown in Figure 9.

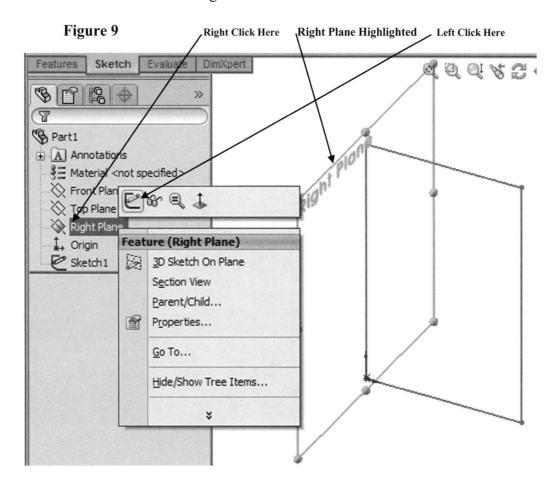

Figure 9

Chapter 8: Introduction to Advanced Commands

12. Your screen should look similar to Figure 10.

Figure 10

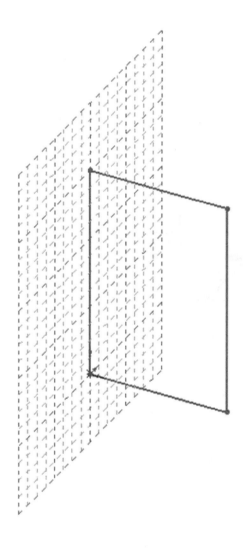

13. Complete the following sketch. Use the Rotate command to provide a better view of the sweep line. The angle of the lines can be estimated. **The sketch lines must intersect with the corner of the 2 inch by 3 inch box as shown in Figure 11**. To dimension on an angle keep the cursor close to the line to be dimensions (while the dimension is still attached to the cursor) then left click once. Press the **Esc** key on the keyboard after completing the dimensioning.

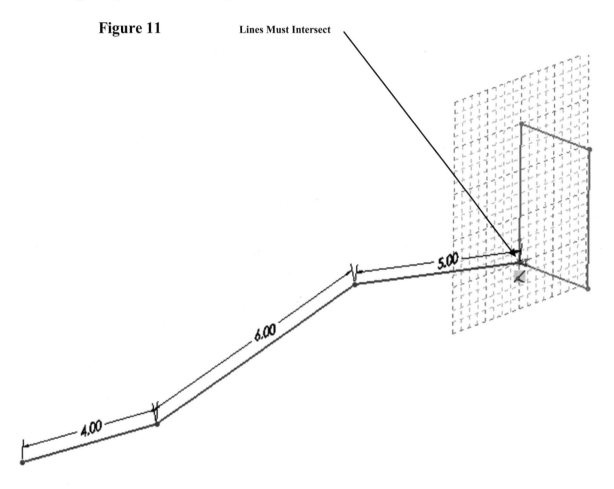

Figure 11   Lines Must Intersect

Chapter 8: Introduction to Advanced Commands

14. Right click anywhere on the screen. A pop up menu will appear. Left click on **Exit Sketch** as shown in Figure 12.

Figure 12

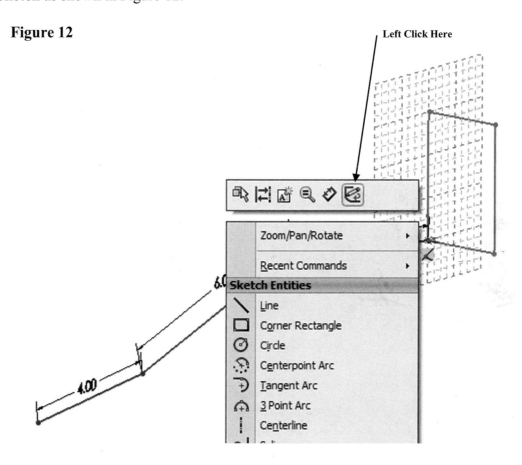

390

Chapter 8: Introduction to Advanced Commands

15. Your screen should look similar to Figure 13.

Figure 13

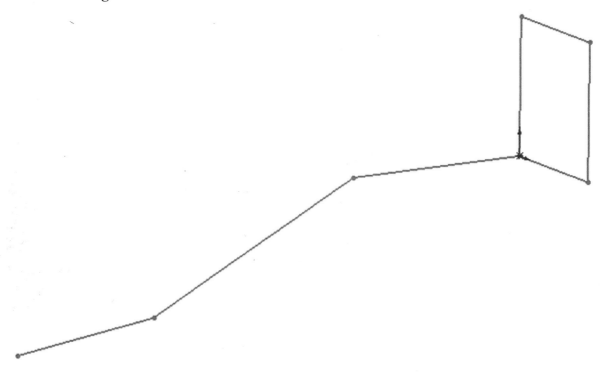

16. Move the cursor to the upper left portion of the screen and left click on **Swept Boss/Base** as shown in Figure 14.

Figure 14

Left Click Here

Chapter 8: Introduction to Advanced Commands

17. Left click on one of the lines that make up the rectangle as shown in Figure 15.

Figure 15

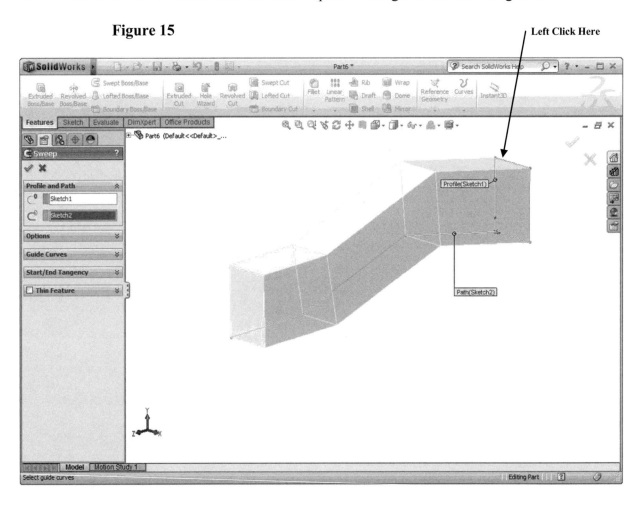

Left Click Here

Chapter 8: Introduction to Advanced Commands

18. SolidWorks will provide a preview of the sweep. Left click on the green check mark as shown in Figure 16.

Figure 16

19. Your screen should look similar to Figure 17.

Figure 17

393

Chapter 8: Introduction to Advanced Commands

20. Move the cursor to the upper middle portion of the screen and left click on **Shell** as shown in Figure 18.

Figure 18

Left Click Here

21. Move the cursor to the left side face and left click once. Using the **Rotate** command, rotate the part around to gain access to the right side <u>face</u> and left click once as shown in Figure 19.

Figure 19

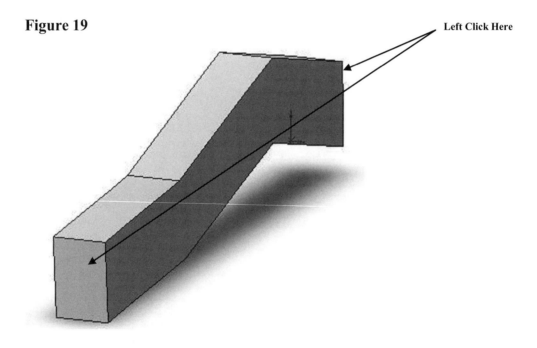

Left Click Here

394

22. Left click on the green check mark as shown in Figure 20.

Figure 20

23. Use the **Rotate** command to access the ends of the model. The model should be open on both ends similar to a piece of rectangular tubing. Your screen should look similar to Figure 21.

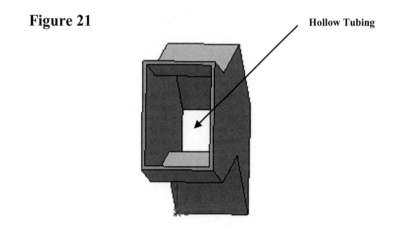

Figure 21

Chapter 8: Introduction to Advanced Commands

24. Begin a New Sketch.

25. Move the cursor to the upper left portion of the screen and left click on **Rectangle** as shown in Figure 22.

Figure 22

26. Move the cursor to the center of the screen and left click once. Ensure that the red dot appears on the intersection of the origin as shown in Figure 23.

Figure 23

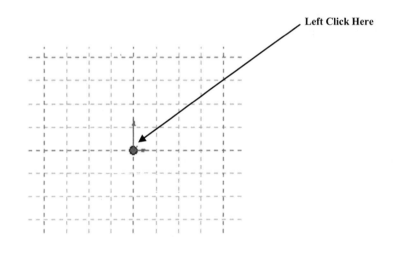

396

Chapter 8: Introduction to Advanced Commands

27. Complete the sketch as shown in Figure 24. Right click anywhere on the screen. A pop up menu will appear. Left click on **Select** as shown in Figure 24.

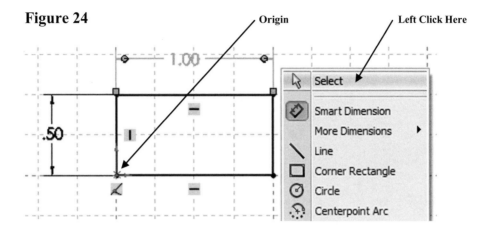

Figure 24

28. Right click anywhere on the screen. A pop up menu will appear. Left click on **Exit Sketch** as shown in Figure 25.

Figure 25

Chapter 8: Introduction to Advanced Commands

29. Move the cursor to the upper middle portion of the screen and left click on the drop down arrow to the right of the Views Orientation icon. A drop down menu will appear. Left click on **Trimetric** as shown in Figure 26.

    Figure 26

    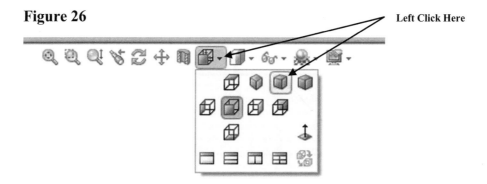

30. Use the **Extruded Boss/Base** command to extrude the sketch to a thickness of .001 of an inch as shown in Figure 27.

    Figure 27

    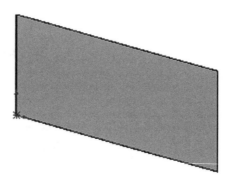

Chapter 8: Introduction to Advanced Commands

31. Move the cursor to the upper middle portion of the screen and left click on the drop down arrow underneath Reference Geometry. A drop down menu will appear. Left click on **Plane** as shown in Figure 28.

Figure 28

399

Chapter 8: Introduction to Advanced Commands

32. Left click on the surface of the solid. SolidWorks will provide a preview of the new plane. Left click on the Distance icon and enter 1.00. Left click on the green check mark as shown in Figure 29.

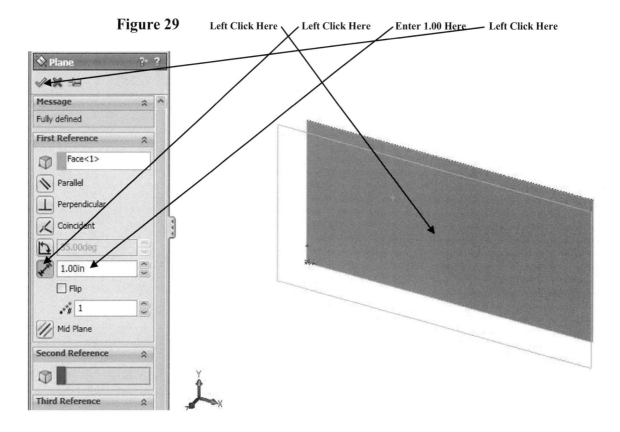

Figure 29

Chapter 8: Introduction to Advanced Commands

33. Your screen should look similar to Figure 30.

Figure 30

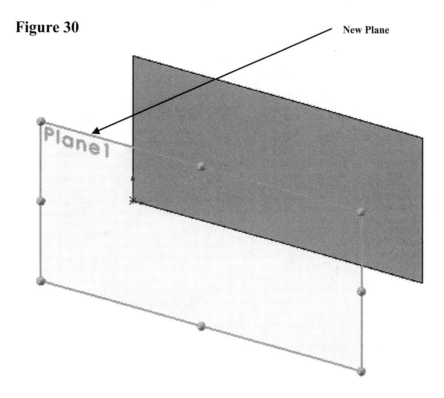

34. Move the cursor to the edge of the newly created plane and right click. A pop up menu will appear. Left click on **Sketch** as shown in Figure 31.

Figure 31

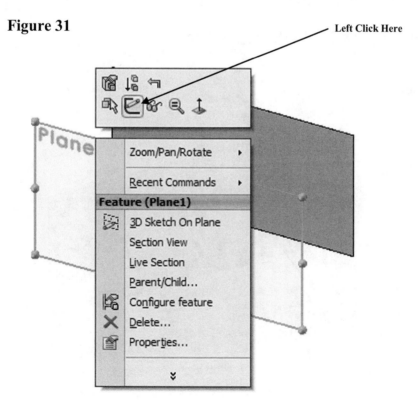

401

Chapter 8: Introduction to Advanced Commands

35. A new sketch will appear on the newly created plane as shown in Figure 32.

Figure 32

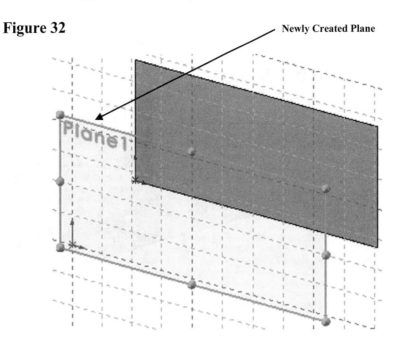

36. Draw a small rectangle in the corner of the sketch as shown in Figure 33. Estimate size and distance.

Figure 33

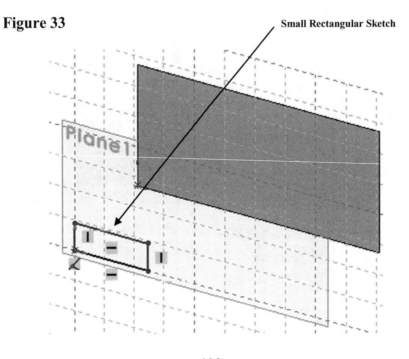

Chapter 8: Introduction to Advanced Commands

37. Once the sketch is complete, right click anywhere around the part. A pop up menu will appear. Left click on **Exit Sketch** as shown in Figure 34.

Figure 34

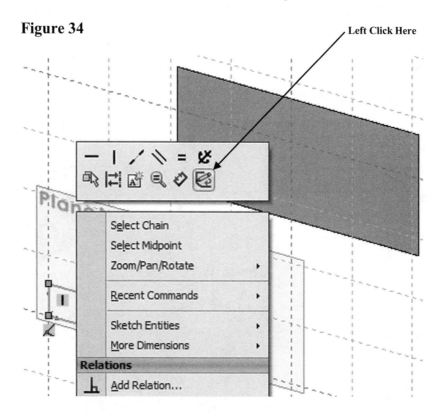

38. Move the cursor to the upper middle portion of the screen and left click on the drop down arrow underneath Reference Geometry. A drop down menu will appear. Left click on **Plane** as shown in Figure 35.

Figure 35

403

Chapter 8: Introduction to Advanced Commands

39. Left click on the previously create workplane. SolidWorks will place another plane 1.00 inch in front of the previous plane. If another plane does not appear, left click on the edge of Plane 1. Left click on the green check mark. Your screen should look similar to Figure 36.

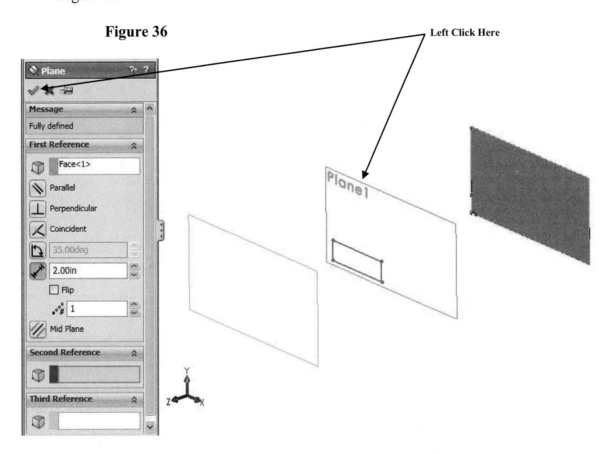

Figure 36

40. Your screen should look similar to Figure 37.

Figure 37

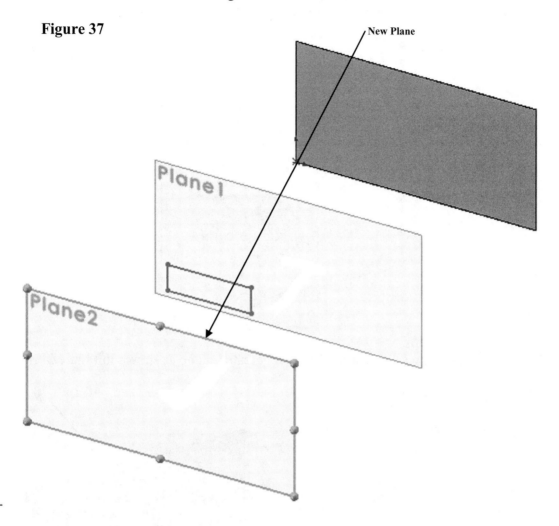

Chapter 8: Introduction to Advanced Commands

41 Move the cursor to the edge of the newly created plane and right click. A pop up menu will appear. Left click on **Sketch** as shown in Figure 38.

Figure 38

42. A new sketch will appear on the newly created plane as shown in Figure 39.

Figure 39

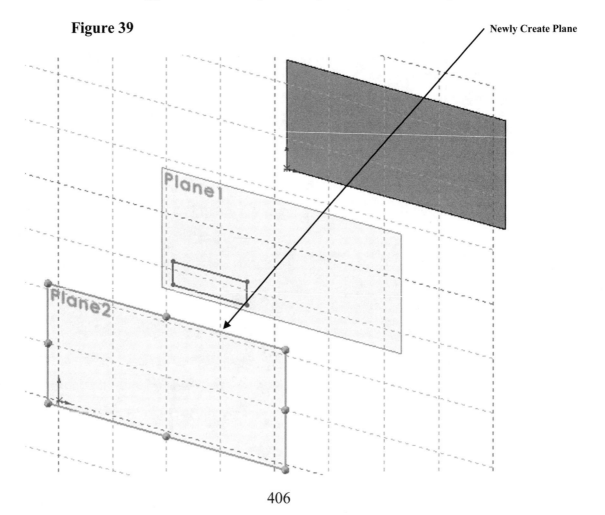

Chapter 8: Introduction to Advanced Commands

43. Draw a small circle in the corner of the sketch as shown in Figure 40. Estimate size and distance.

Figure 40

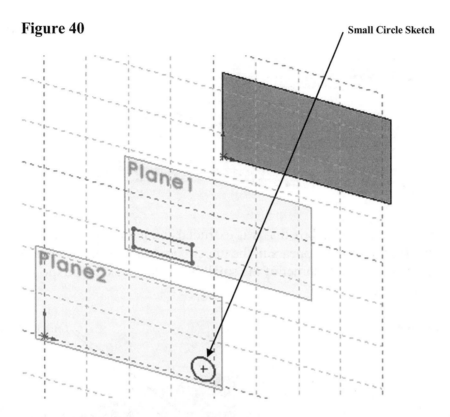

Small Circle Sketch

44. Once the sketch is complete, right click anywhere around the part. A pop up menu will appear. Left click on **Exit Sketch** as shown in Figure 41.

Figure 41

Left Click Here

407

Chapter 8: Introduction to Advanced Commands

45. Move the cursor to the upper left portion of the screen and left click on **Lofted Boss/Base** as shown in Figure 42.

Figure 42

Left Click Here

46. Move the cursor to the first box that was extruded and left click on the face. Because the Profile banner appears in the upper corner of the first sketch, move the cursor to the upper left corner of the small box (second sketch in the same location) and left click.

Figure 43

Left Click Here

47. Your screen should look similar to Figure 44.

Figure 44

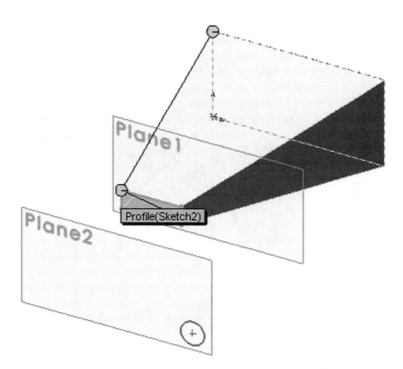

48. Move the cursor to the <u>edge</u> of the circle and left click as shown in Figure 45.

Figure 45

49. Your screen should look similar to Figure 46.

Figure 46

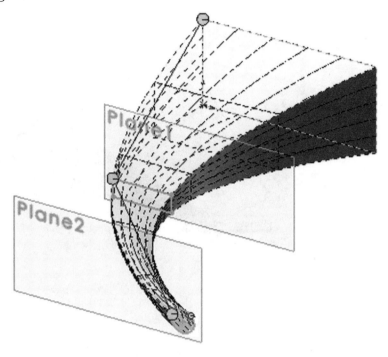

50. Left click on the green check mark as shown in Figure 47. Press the **Esc** key several times.

Figure 47

51. Your screen should look similar to Figure 48.

Figure 48

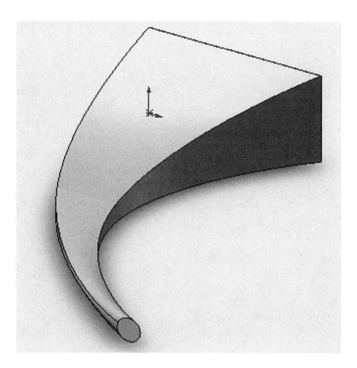

Chapter 9 Introduction to SimulationXpress

Objectives:
- Learn to create simple parts used to run SimulationXpress
- Learn to interpret the results of an analysis

Chapter 9 includes instruction on how to run the analysis shown below.

Chapter 9: Introduction to SimulationXpress

1. Start SolidWorks by referring to "Chapter 1 Getting Started".

2. Begin a new Sketch.

3. Complete the sketch as shown in Figure 1.

   Figure 1

4. Extrude the sketch a distance of 2 inches as shown in Figure 2.

   Figure 2

Chapter 9: Introduction to SimulationXpress

5. Save the part where it can easily be retrieved later.

6. Move the cursor to the upper left portion of the screen and left click on the arrow to the right of SolidWorks. A drop down menu will appear. Left click on **SimulationXpress** as shown in Figure 3.

Figure 3

Chapter 9: Introduction to SimulationXpress

7. The SimulationXpress dialog box will appear. Left click on **Next** as shown in Figure 4.

Figure 4

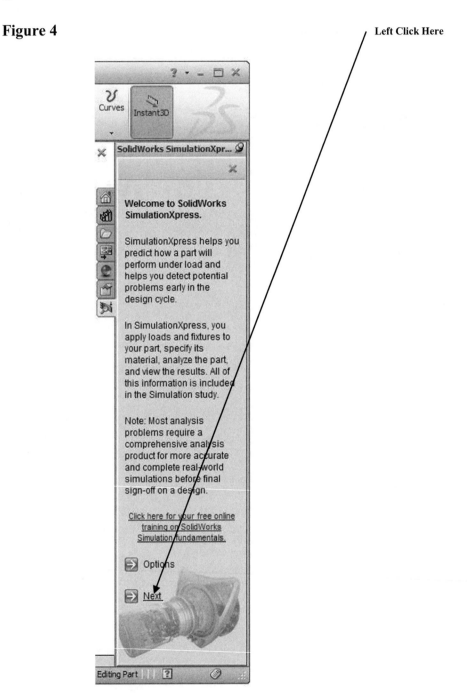

Left Click Here

416

Chapter 9: Introduction to SimulationXpress

8. Left click on **Add a Fixture** as shown in Figure 5.

Figure 5

Left Click Here

Chapter 9: Introduction to SimulationXpress

9. Use the **Rotate** command to rotate the part upward to gain access to the bottom as shown in Figure 6.

Figure 6

10. Left click on the bottom surfaces as shown in Figure 7.

Figure 7

Left Click Here

Chapter 9: Introduction to SimulationXpress

11. Move the cursor to the upper left portion of the screen and left click on the green check mark as shown in Figure 8.

Figure 8

Left Click Here

12. Move the cursor to the upper right portion of the screen and left click on **Loads** as shown in Figure 9.

Figure 9

Left Click Here

419

Chapter 9: Introduction to SimulationXpress

13. Use the **Rotate** command to rotate the part downward to gain access to the top as shown in Figure 10.

Figure 10

14. Left click on the top of the part as shown in Figure 11.

Figure 11

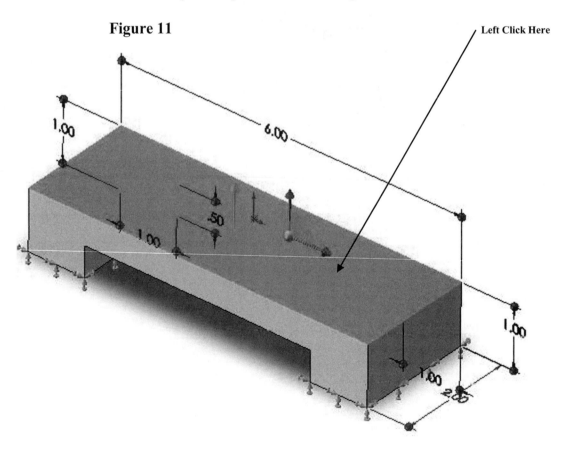

Chapter 9: Introduction to SimulationXpress

15. Move the cursor to the lower right portion of the screen and left click on **Add a force** as shown in Figure 12.

Figure 12

Left Click Here

Chapter 9: Introduction to SimulationXpress

16. Move the cursor to the middle left portion of the screen and left click on the drop down arrow and change the units to **English (IPS)** as shown in Figure 13.

Figure 13

17. Enter **100** for the load value and left click on the green check mark as shown in Figure 14

Figure 14

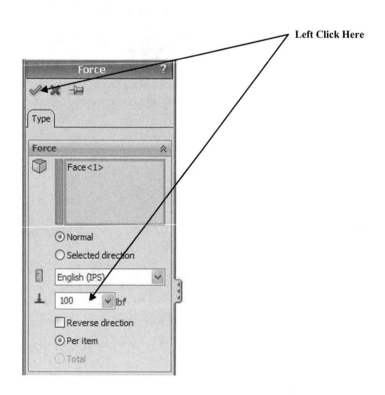

422

Chapter 9: Introduction to SimulationXpress

18. Move the cursor to the upper right portion of the screen and left click on **Material.** Left click on **Choose Material** as shown in Figure 15

Figure 15   Left Click Here

Chapter 9: Introduction to SimulationXpress

19. The Material dialog box will appear. Scroll down to **Aluminum Alloys** and select **1060 Alloy**. Left click on **Apply**. Left click on **Close** as shown in Figure 16.

Figure 16

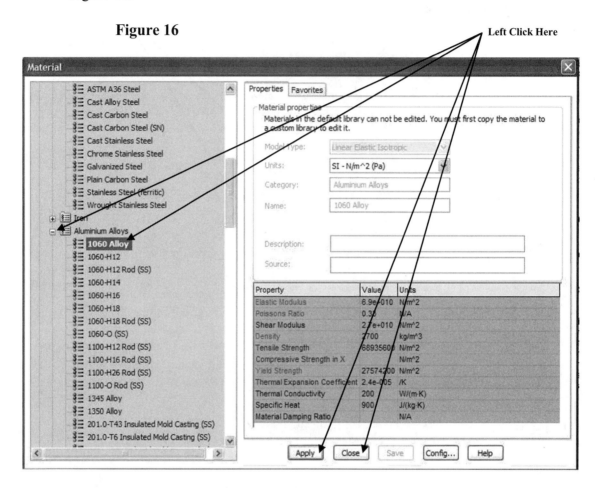

Chapter 9: Introduction to SimulationXpress

20. Move the cursor to the upper right portion of the screen and left click on **Run.** Left click on **<u>Run Simulation</u>** as shown in Figure 17.

Figure 17

Chapter 9: Introduction to SimulationXpress

21. The model will go through the simulation process. Notice that SolidWorks has moved to the Results portion of the simulation. Once this is complete, SolidWorks will display the results. Left click on **Yes, continue** as shown in Figure 18.

Figure 18

Results Portion   Left Click Here

426

Chapter 9: Introduction to SimulationXpress

22. Move the cursor to the upper right portion of the screen and left click on **Show von Mises** stress. SolidWorks will provide an exaggerated display of the simulation during the test. Your screen should look similar to Figure 19. Rotate the part upward to identify any problems areas in the design. Areas of red indicate a potential high stress area. Left click on **Play Animation** to watch the simulation with color representation. When you are finished viewing the animation, move the cursor to the lower right portion of the screen and left click on **Stop Animation**. Move the cursor to the upper right corner of the screen and left click on **Close** as shown in Figure 19.

Figure 19

23. Begin a new drawing. Complete the sketch shown in Figure 20.

Figure 20

76,2

50,8

24. Extrude the sketch to a distance of **96** inches as shown in Figure 21.

2438 mm

Figure 21

25. Use the **Shell** command to shell the part with a thickness of **.10** as shown in Figure 22.

Figure 22

26. Move the cursor to the upper left portion of the screen and left click on the arrow to the right of SolidWorks. A drop down menu will appear. Left click on **SimulationXpress** as shown in Figure 23.

Figure 23

Chapter 9: Introduction to SimulationXpress

27. Run a simulation on this part (using **Cast Alloy Steel** for material and **100 lbs** for the Load) using the left side as the Restraint surface and the top as the Load surface as shown in Figure 24. This will create a cantilever effect for the simulation. If the simulation requires too much memory to run, reduce the length of the lever and run the simulation again.

Figure 24 — Restraint Surface / Load Surface

28. After the simulation has been run, your screen should look similar to Figure 25.

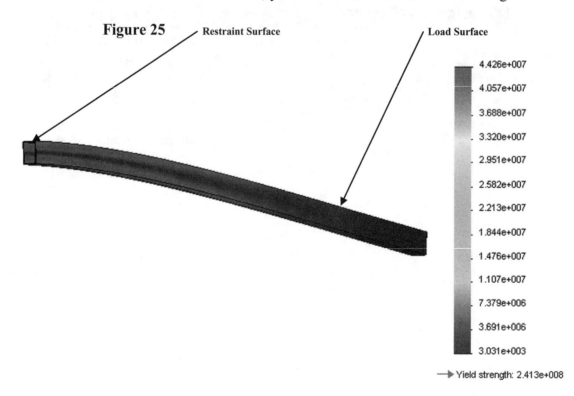

Figure 25 — Restraint Surface / Load Surface

Chapter 10 Introduction to Mechanical Mates

Objectives:
- Learn to create a cam and lifter system of parts
- Learn to use one of the Mechanical Mates options

Chapter 10 includes instruction on how to construct the assembly shown below.

Chapter 10: Introduction to Mechanical Mates

1. Start SolidWorks by referring to "Chapter 1 Getting Started".

2. After SolidWorks is running, begin a New Sketch.

Figure 1

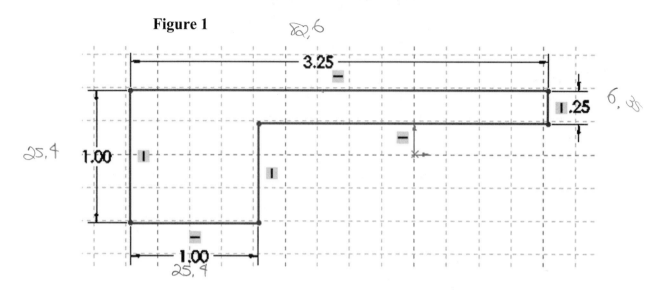

3. Exit the Sketch area. Extrude the sketch to a thickness of **1.00** in as shown in Figure 2

Figure 2

4. Use the **Fillet** command (.5 inch fillets) to radius the lower edge(s) as shown in Figure 3.

Figure 3

5. Use the **Rotate** command to rotate the part upward as shown in Figure 4.

Figure 4

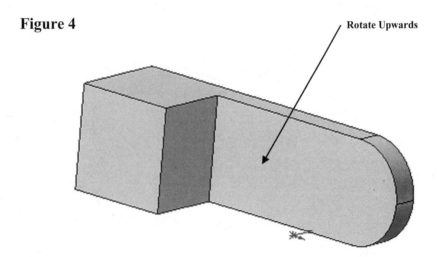

Rotate Upwards

6. Complete the sketch as shown in Figure 5. The center of the circle is located on the corner of the fillet radius.

Figure 5

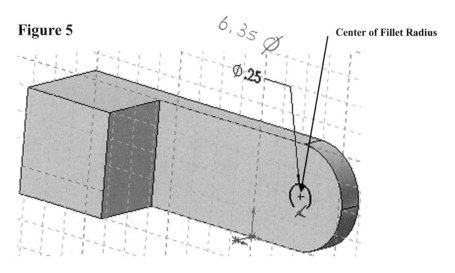

7. Extrude the circle to a distance of **.75** inches as shown in Figure 6.

Figure 6

8. Rotate the part around. Complete the sketch shown in Figure 7.

Figure 7

9. Use the **Extruded Cut** option to cut a hole a distance of **1.00 inch** as shown in Figure 8.

Figure 8 — Hole Cut 1.00 Inch

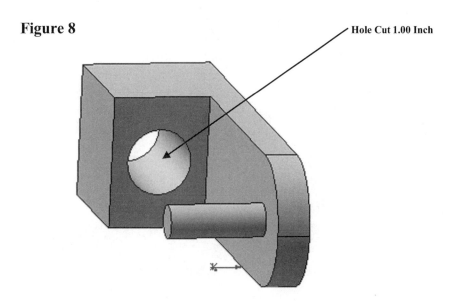

Chapter 10: Introduction to Mechanical Mates

10. Save the part as **Camcase1.SLDPRT** where it can be easily retrieved later.

11. Complete the sketch shown. Extrude the .500 in diameter circle to a distance of 1.00 inch and the .625 in diameter circle to a distance of .125 inches as shown in Figure 9.

Figure 9

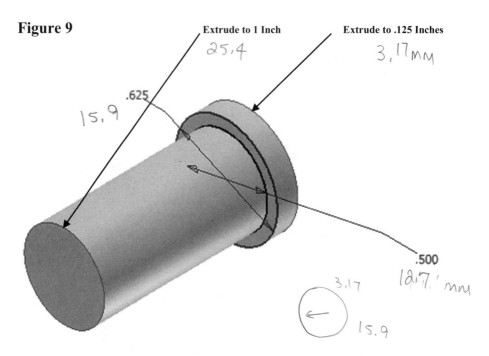

12. The bottom of the part needs to be flat as shown in Figure 10.

Figure 10

13. Save the part as **Lifter1.SLDPRT** where it can easily be retrieved later.

14. Begin a new drawing. Complete the sketch shown in Figure 11.

Figure 11

Chapter 10: Introduction to Mechanical Mates

15. Extrude the sketch to a distance of **.500** inches as shown in Figure 12.

Figure 12 12.7mm

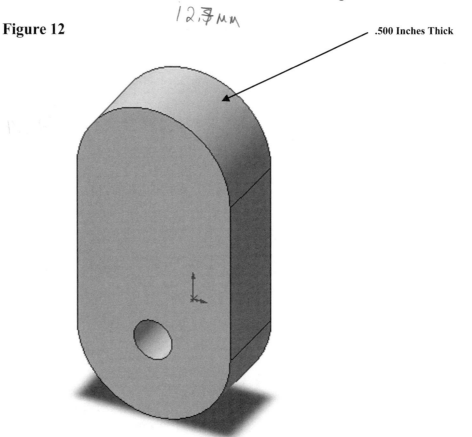

.500 Inches Thick

16. Save the part as **Lobe1.SLDPRT** where it can easily be retrieved later.

17. Start a new Assembly drawing as shown in Chapter 7.

Chapter 10: Introduction to Mechanical Mates

18. Move the cursor to the lower middle portion of the screen and left click on **Browse**. The Open dialog box will appear. Left click on **Camcase1.SLDPRT**. Left click on **Open** as shown in Figure 13.

Figure 13

19. SolidWorks will place one cam case in the work area. The cam case will be attached to the cursor. If the part does not appear attached to the cursor simply left click anywhere in the work area.

20. Use the **Insert Components** command to insert the **Lifter1.SLDPRT** and **Lobe1.SLDPRT** parts as shown in Figure 14.

Figure 14

439

Chapter 10: Introduction to Mechanical Mates

21. Your screen should look similar to Figure 15.

Figure 15

22. Using the **Mate** command, place the lifter in the lifter bore as shown. You may have to change the Mate Alignment, located at the lower left portion of the screen if the lifter is mated upside down as shown in Figure 16.

Figure 16   Left Click Here If Lifter Mates Upside Down   Lifter Foot

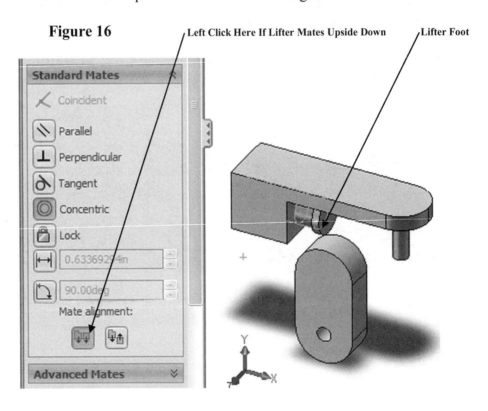

440

23. Rotate the assembly up as shown in Figure 17.

Figure 17

24. Use the **Mate** command to mate the lobe to the shaft as shown in Figure 18.

Figure 18

Chapter 10: Introduction to Mechanical Mates

25. Use the **Mate** command to create a **.125 inch** offset between the lobe and the case as shown in Figure 19.

Figure 19

.125 Inch Offset

26. Use the **Rotate** command to rotate the assembly around as shown in Figure 20.

Figure 20

Chapter 10: Introduction to Mechanical Mates

27. Move the cursor to the upper left portion of the screen and left click **Mate** as shown in Figure 21.

Figure 21

28. Left click on drop down arrows to the right of Mechanical Mates as shown in Figure 22.

Figure 22

443

Chapter 10: Introduction to Mechanical Mates

29. The Mechanical Mates drop down box will appear.

30. Left click on **Cam** as shown in Figure 23.

Figure 23

31. Left click on all 4 faces of the cam. You will have to rotate the assembly around (by pressing down on the mouse wheel) to gain access to all 4 faces as shown in Figure 24.

Figure 24

444

Chapter 10: Introduction to Mechanical Mates

32. Left click in the box under Cam Follower as shown in Figure 25.

Figure 25

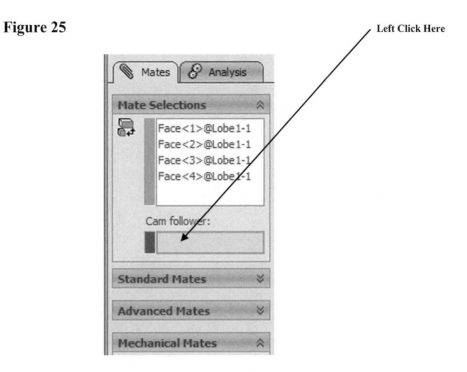

33. Left click on the Lifter foot as shown in Figure 26.

Figure 26

Chapter 10: Introduction to Mechanical Mates

34. SolidWorks will move the lifter down in contact with the cam face as shown in Figure 27.

Figure 27

35. Left click on the green check mark as shown in Figure 28.

Figure 28

36. An error message will appear indicating that portions of the lobe contain a flat surface. This was done to simplify the exercise. Left click on **OK** as shown in Figure 29.

Figure 29

37. Move the cursor to the upper left portion of the screen and left click on the green check mark as shown in Figure 30.

Figure 30

38. Move the cursor to the upper middle portion of the screen and left click on **New Motion Study** as shown in Figure 31.

Figure 31

Chapter 10: Introduction to Mechanical Mates

39. Left click on the **Motor** icon as shown in Figure 32.

Figure 32

Left Click Here

40. Left click on the lower portion of the lobe (base circle) as shown in Figure 33.

Figure 33

Left Click Here

Chapter 11 Advanced Work Plane Procedures

Objectives:
- Learn to create points on a solid model
- Learn to create an offset work plane
- Learn to create an offset Extruded cut

Chapter 11 includes instruction on how to create multiple Points on a solid model and use these points to create an offset work plane for use as an offset cut extrusion.

Chapter 11 includes instruction on how to design the part shown below.

Chapter 11: Advanced Work Plane Procedures

1. Start SolidWorks by referring to "Chapter 1 Getting Started". After SolidWorks is running, begin a New Sketch.

2. Complete the sketch shown in Figure 1.

Figure 1

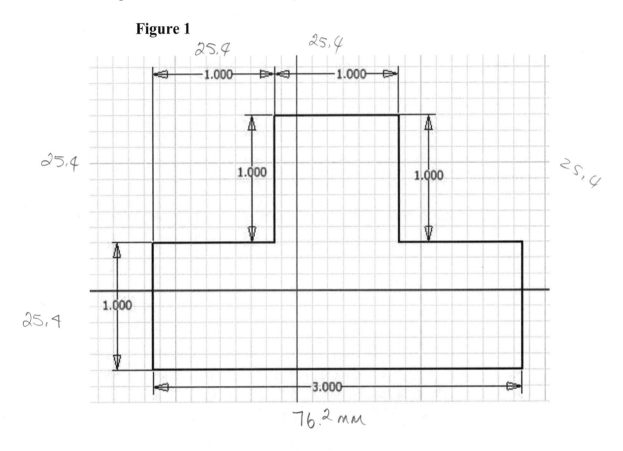

Chapter 11: Advanced Work Plane Procedures

3. Exit the Sketch Panel. Extrude the sketch to a thickness of **4.00** inches as shown in Figure 2.

Figure 2

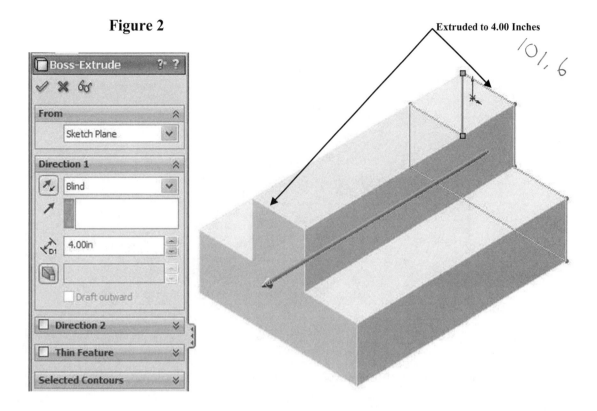

4. Your screen should look similar to Figure 3

Figure 3

Chapter 11: Advanced Work Plane Procedures

5. Use the **Rotate** command to rotate the part to gain a perpendicular view of the front surface as shown in Figure 4.

   Figure 4

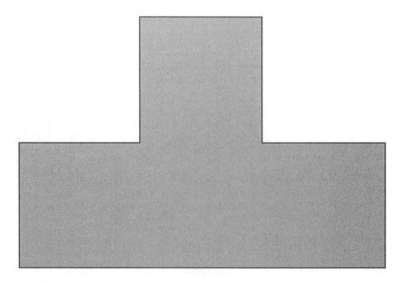

6. Use the **Rectangle** command to complete the sketch as shown in Figure 5. All dimensions are .188.

   Figure 5

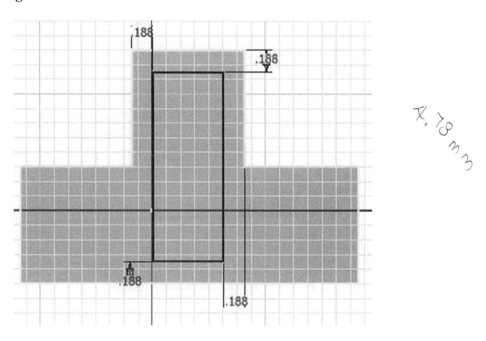

Chapter 11: Advanced Work Plane Procedures

7. Extrude (cut) the rectangle a distance of **4.00** inches creating a square hole as shown in Figure 6.

Figure 6

8. Rotate the part around to gain an isometric (Home View) of the part as shown in Figure 7.

Figure 7

Chapter 11: Advanced Work Plane Procedures

9. Rotate the part around to gain an isometric view of the side of the part as shown in Figure 8.

Figure 8

10. Begin a **New Sketch** on the front surface of the part as shown in Figure 9.

Figure 9

Begin a New Sketch on This Surface

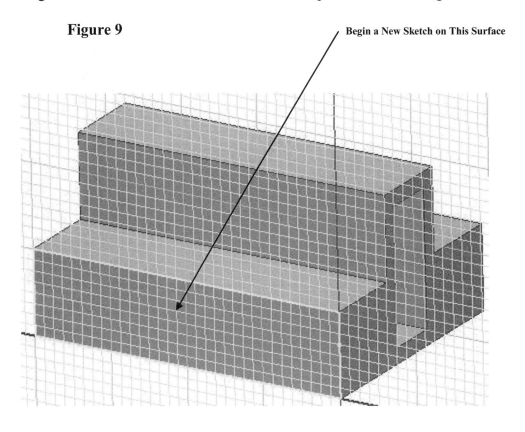

11. Complete the sketch (2 points) as shown in Figure 10.

Figure 10

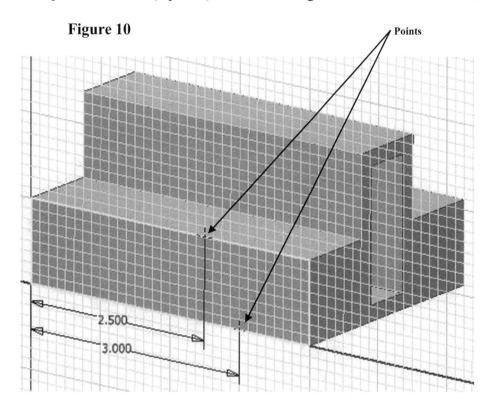

12. Exit out of the Sketch as shown in Figure 11.

Figure 11

Chapter 11: Advanced Work Plane Procedures

13.   Rotate the part upward as shown in Figure 12.

Figure 12

Chapter 11: Advanced Work Plane Procedures

14. Begin a **New Sketch** on the surface shown in Figure 13.

Figure 13

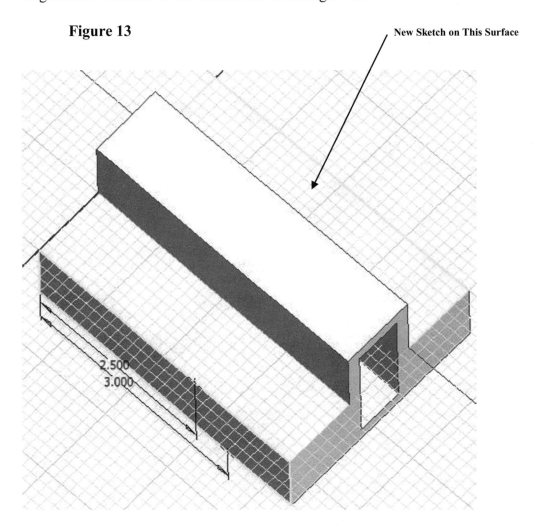

15. Create a **Point** on the surface as shown in Figure 14.

Figure 14

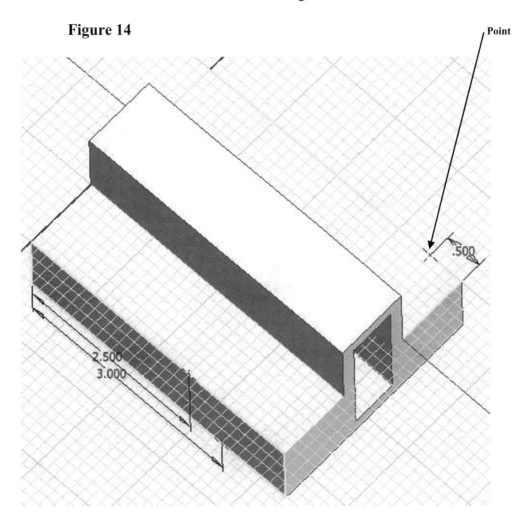

Chapter 11: Advanced Work Plane Procedures

16. Exit the Sketch as shown in Figure 15.

Figure 15

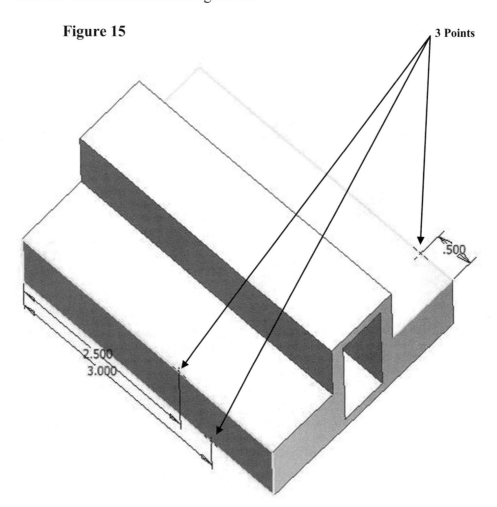

Chapter 11: Advanced Work Plane Procedures

17. Move the cursor to the upper middle portion of the screen and left click on the arrow below Reference Geometry. A drop down menu will appear. Left click on **Plane** as shown in Figure 16.

Figure 16

18. Left click on each of the 3 points. The dimensions are included in the figure below for instructional purposes. They may not be visible as shown in Figures 17-23.

Figure 17

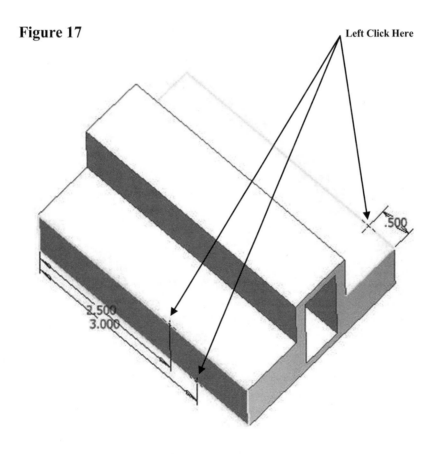

463

Chapter 11: Advanced Work Plane Procedures

19. SolidWorks will create a Work Plane from the 3 points. Your screen should look <u>similar</u> to Figure 18.

Figure 18

20. Rotate the newly created work plane upward perpendicular as shown in Figure 19.

Figure 19

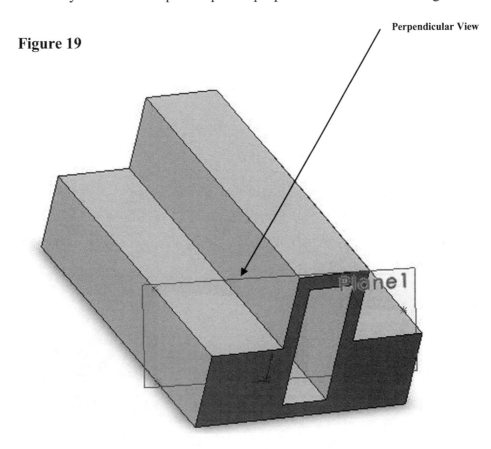

Chapter 11: Advanced Work Plane Procedures

21. Move the cursor over the edge of the newly created plane causing the edges to turn red and create a New Sketch (right click, etc) as shown in Figure 20.

Figure 20

Left Click Here

22. SolidWorks will create a New Sketch on the newly created work plane as shown in Figure 21.

Figure 21

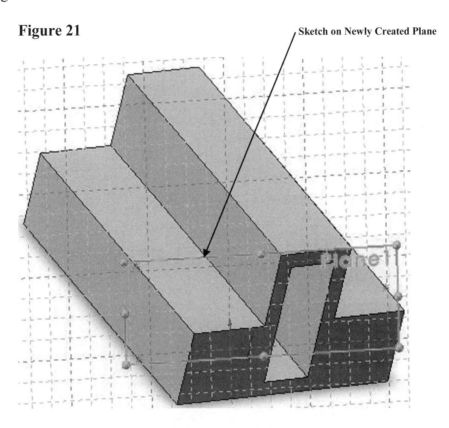

Sketch on Newly Created Plane

Chapter 11: Advanced Work Plane Procedures

23. Draw a large rectangle that extends out past all for edges of the part as shown in Figure 22.

Figure 22

24. Exit the Sketch and use the **Extruded Cut** command to create a removal portion similar to Figure 23.

Figure 23

Chapter 11: Advanced Work Plane Procedures

25. Your screen should look similar to Figure 24.

Figure 24

Chapter 11: Advanced Work Plane Procedures

26. To clean up the appearance of the part, move the cursor over the left portion of the screen and left click, then right click on **Plane1**. A pop up menu will appear. Left click on the glasses (the hide command) as shown in Figure 25.

Figure 25

Chapter 11: Advanced Work Plane Procedures

27. Your screen should look similar to Figure 26.

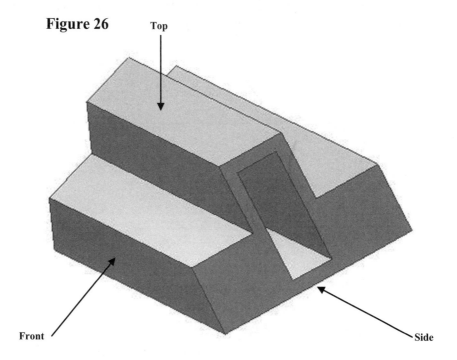

Figure 26

28. To test your Orthographic skills, create a 3 view Orthographic drawing of the part shown above (Top, Front and Right Hand side views) in either a 2 Dimensional CAD package or by hand on a Drafting Board. Show all hidden lines in your 2 Dimensional drawing. Once you have completed your 2 Dimensional drawing, refer to Chapter 3 on how to create a 3 View drawing in SolidWorks, (showing all hidden lines) and then compare your "hand drawing" to the SolidWorks 3 view drawing.

Chapter 12 Introduction to Creating a Helix

Objectives:
- Learn to create a Helix

Chapter 12 includes instruction on how to create a Helix

Chapter 12 includes instruction on how to design the part shown below.

Chapter 12: Introduction to Creating a Helix

1. Start SolidWorks by referring to "Chapter 1 Getting Started". After SolidWorks is running, begin a New Sketch.

2. Complete the sketch shown in Figure 1.

   Figure 1

3. Rotate the drawing around as shown in Figure 2.

   Figure 2

   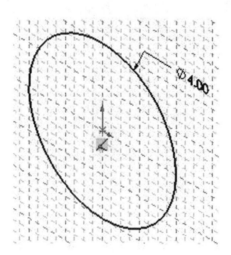

Chapter 12: Introduction to Creating a Helix

4. Exit the Sketch Panel. Your screen should look similar to Figure 3

Figure 3

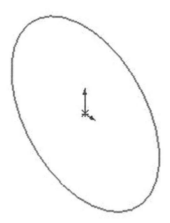

5. Move the cursor to the upper middle portion of the screen and left click on the arrow below Curves. A drop down menu will appear. Left click on **Helix and Spiral** as shown in Figure 4.

Figure 4

475

Chapter 12: Introduction to Creating a Helix

6. Left click on the circle as shown in Figure 5. SolidWorks will provide a preview of the Helix.

Figure 5

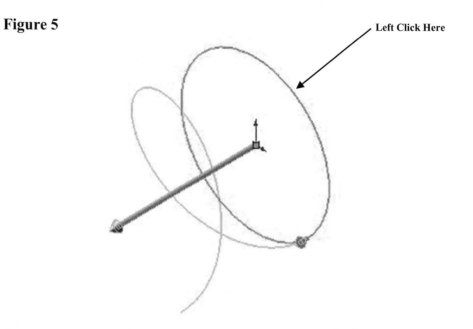

7. Enter **2.00** for the Pitch and **10** for the Revolutions. Left click on the green check mark as shown in Figure 6.

Figure 6

476

Chapter 12: Introduction to Creating a Helix

8. Your screen should look similar to Figure 7.

Figure 7

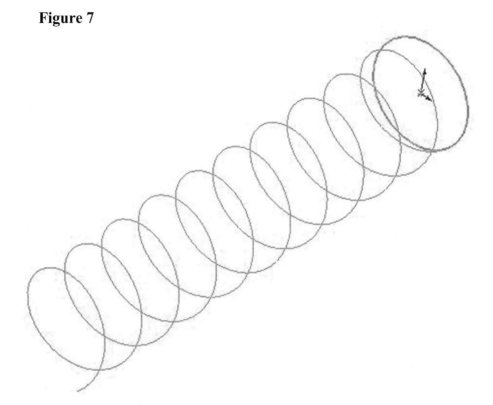

9. Move the cursor to the upper left portion of the screen and begin a **New Sketch** on the **Right Plane** as shown in Figure 8.

Figure 8

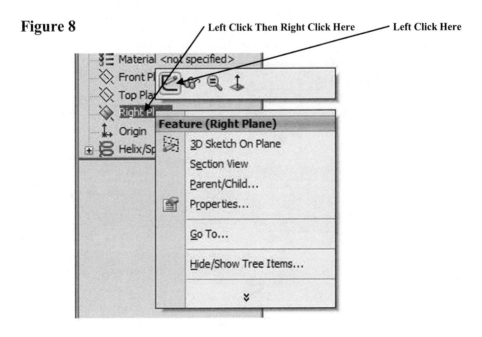

477

Chapter 12: Introduction to Creating a Helix

10. Rotate the view slightly Isometric. Left click on the Helix as shown in Figure 9.

Figure 9

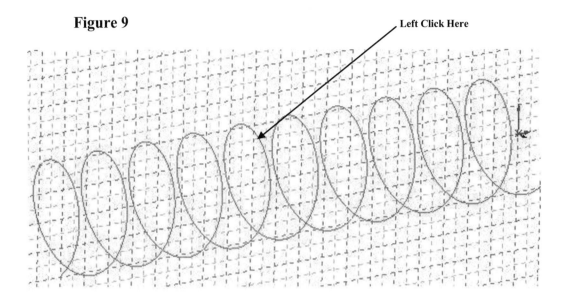

Left Click Here

11. Move the cursor to the upper middle portion of the screen and left click on **Convert Entities** as shown in Figure 10.

Figure 10

Left Click Here

478

12. SolidWorks will convert the Helix into a 2 dimensional drawing as shown in Figure 11.

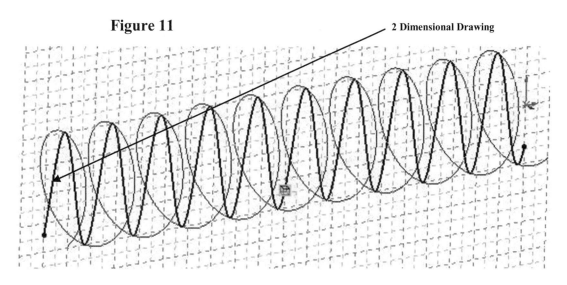

Figure 11 — 2 Dimensional Drawing

13. Draw a 1.00 inch circle at the 2 Dimensional end point of the Helix as shown in Figure 12.

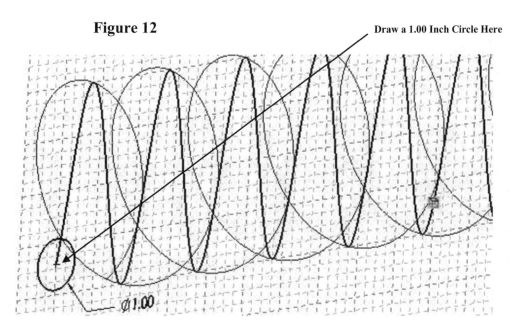

Figure 12 — Draw a 1.00 Inch Circle Here

Chapter 12: Introduction to Creating a Helix

14. Delete the 2 Dimensional Helix drawing of the spring as shown in Figure 13.

Figure 13

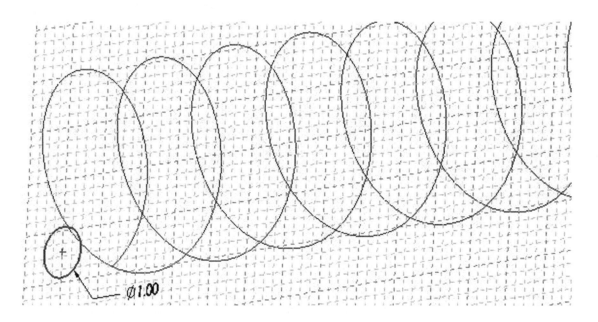

15. Move the cursor to the upper left portion of the screen and left click on **Swept Boss/Base** as shown in Figure 14.

Figure 14

Chapter 12: Introduction to Creating a Helix

16. Left click on the small circle first. Then left click on the Helix as shown in Figure 15.

Figure 15

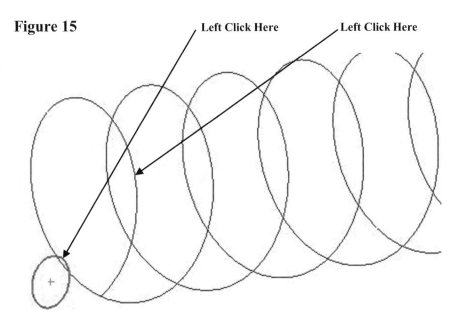

17. SolidWorks will provide a preview of the Helix as shown in Figure 16. Left click on the green check mark (not shown).

Figure 16

481

Chapter 12: Introduction to Creating a Helix

18. Your screen should look similar to Figure 17.

Figure 17

Chapter 13 Introduction to Importing DWG files

Objectives:
- Learn to create a simple .DWG file
- Learn to import a .DWG file into a 1 view drawing
- Learn to import a .DWG file into the solid model area and create a solid model

Chapter 13 includes instruction on how to import a .DWG file into either the solid model area or the orthographic drawing area of SolidWorks.

Chapter 13 includes instruction on how to design the part shown below.

Chapter 13: Introduction to Importing .DWG files

1. Create the following drawing in some type of .DWG editing software (software that is used to create and edit .DWG files such as AutoCAD, DWG Editor, etc) as shown in Figure 1.

   Figure 1

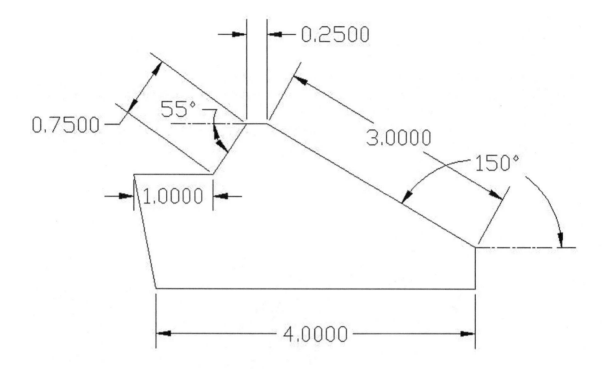

2. Save the drawing where it can be retrieved later. Start SolidWorks by referring to "Chapter 1 Getting Started" with the exception of selecting the "New" icon in the upper left corner of the screen (do not select the "New" icon as in previous chapters). Your screen should look similar to Figure 2.

Figure 2

Chapter 13: Introduction to Importing .DWG files

3. Left click on the "Open" icon as shown in Figure 3.

Figure 3

4. The Open dialog box will appear. Left click on the drop down arrow at the top of the dialog box to locate the drawing file. Left click on the drop down arrow to the right of "Files of Type" to show the different types of files that can be imported. Left click on **.DWG**. The file you just created and saved should appear in the dialog box. Left click on **Open** as shown in Figure 4.

Figure 4

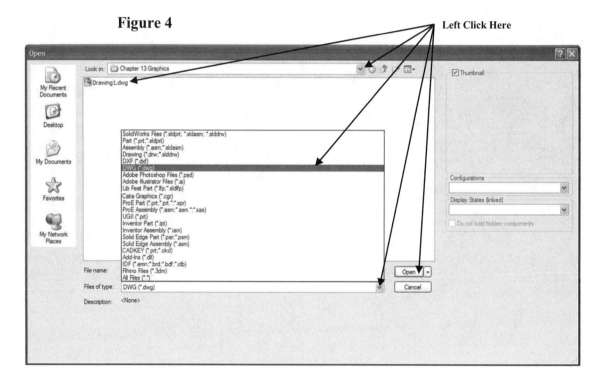

486

Chapter 13: Introduction to Importing .DWG files

5. The DXF/DWG Import dialog box will appear. Left click next to Create new SolidWorks drawing and Convert to SolidWorks entities. Left click on **Next** as shown in Figure 5.

Figure 5

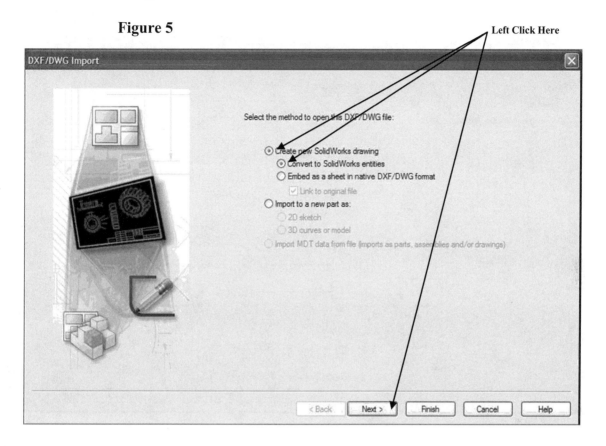

Chapter 13: Introduction to Importing .DWG files

6. The DXF/DWG Import – Drawing Layer Mapping dialog box will appear. Left click on **Next** as shown in Figure 6.

Figure 6

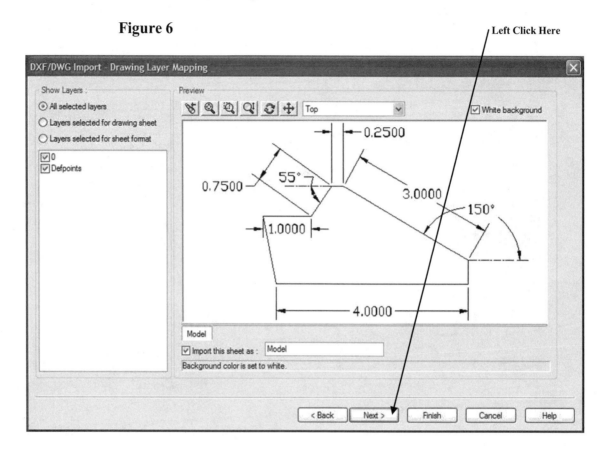

Chapter 13: Introduction to Importing .DWG files

7. The DXF/DWG Import – Document Settings dialog box will appear. Left click on **Finish** as shown in Figure 7.

Figure 7

Chapter 13: Introduction to Importing .DWG files

8. SolidWorks will convert the drawing to a 2 Dimensional one view drawing. The dimensions that were created in the DWG editor are not changeable but portions of the drawing that do not contain dimensions can have dimensions added to them. Create a 100 degree angle dimension as shown in Figure 8.

Figure 8

Create a 100 Degree Angle Dimension Here

Chapter 13: Introduction to Importing .DWG files

9. Enter **100** in the Modify dialog box as shown in Figure 9.

Figure 9

10. Your screen should look similar to Figure 10. Feel free to create other dimensions as desired. Disregard the fact that the addition of the dimension will cause the bottom line to not be parallel.

Figure 10

491

Chapter 13: Introduction to Importing .DWG files

11. Save the drawing where it can be retrieved later. Start SolidWorks by referring to "Chapter 1 Getting Started" with the exception of selecting the "New" icon in the upper left corner of the screen (do not select the "New" icon as in previous chapters). Your screen should look similar to Figure 11.

Figure 11

12. Left click on the **Open** icon as shown in Figure 12.

   Figure 12

13. The Open dialog box will appear. Left click on the drop down arrow at the top of the dialog box to locate the file. Left click on the drop down arrow to the right of Files of Type to show the different types of files that can be imported. Left click on **.DWG**. The file that was originally created and saved should appear in the dialog box. Left click on **Open** as shown in Figure 14.

   Figure 13

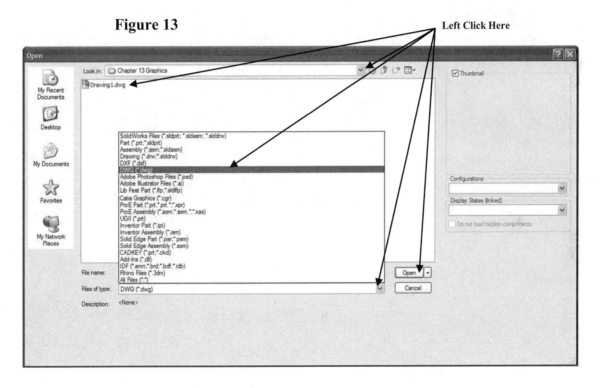

Chapter 13: Introduction to Importing .DWG files

14. The DXF/DWG Import dialog box will appear. Left click next to Import to a new part. Left click on **Next** as shown in Figure 14.

Figure 14

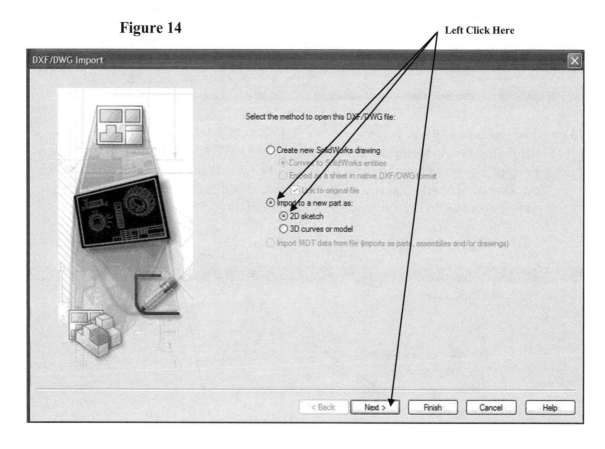

494

Chapter 13: Introduction to Importing .DWG files

15. The DXF/DWG Import – Document Settings dialog box will appear. Left click on **Finish** as shown in Figure 6.

Figure 15

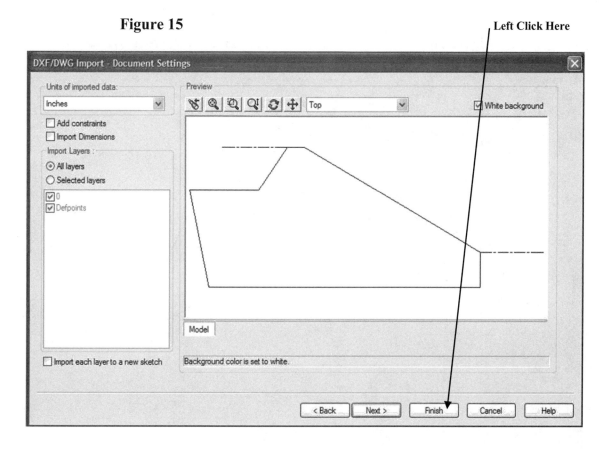

Chapter 13: Introduction to Importing .DWG files

16. SolidWorks will convert the .DWG drawing file into a 2 dimensional sketch that can be dimensioned and extruded into a solid. Delete the 2 construction lines as shown in Figure 16.

Figure 16

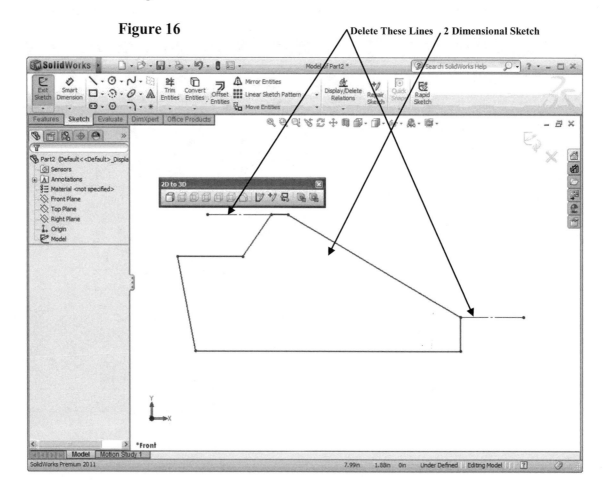

17. Dimension the sketch as shown in Figure 17.

Figure 17

Chapter 13: Introduction to Importing .DWG files

18. Extrude the sketch into a solid with a distance of 2 inches as shown in Figure 18.

Figure 18

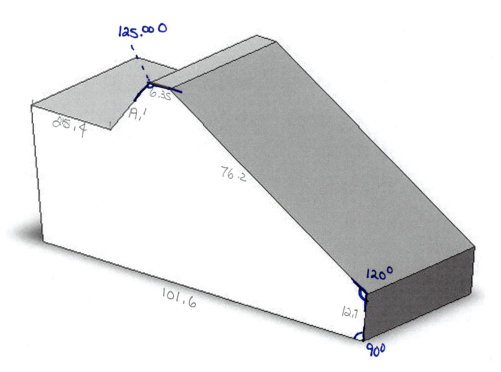

19. Complete the following sketch as shown in Figure 19.

Figure 19

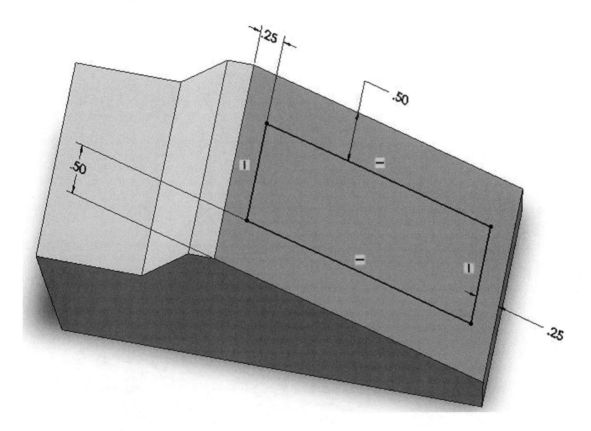

Chapter 13: Introduction to Importing .DWG files

20. Extrude-Cut the rectangle to create a thru hole in the part as shown in Figure 20. You will see that the solid behaves as any solid that was initially created in SolidWorks.

Figure 20

Thru Hole

Chapter 14 Introduction to the Design Library

Objectives:
- Learn to import pre-designed parts from the Design Library Toolbox

Chapter 14 includes instruction on how to import pre-designed parts from the Design Library Toolbox and create a Part drawing file.

Chapter 14 includes instruction on how to use the Design Library Toolbox to design the part shown below.

Chapter 14: Introduction to the Design Library

1. SolidWorks has many pre-designed parts that can be imported to create a Part drawing file. These parts are located in the Design Library Toolbox. The purpose of the Design Library Toolbox is to reduce the amount of time required to design parts that are already available through a part supplier. Common items such as bearings, C-clip's bolts, nuts and washers are typically produced in mass quantity by part suppliers. For this reason designers typically do not design the above mentioned items. They simply import/create them in multiple numbers from the Design Library Toolbox.

2. Start SolidWorks by referring to "Chapter 1 Getting Started". Do NOT begin a New Drawing. You screen should look similar to Figure 1

   Figure 1

Chapter 14: Introduction to the Design Library

3. Move the cursor to the upper right portion of the screen and left click on the **Design Library** icon if the Design Library is not already open as shown in Figure 2.

Figure 2

Chapter 14: Introduction to the Design Library

4. Move the cursor to the upper right portion the screen and left click on **Toolbox**. Move the cursor to the lower right portion of the screen and left click on ***Add in Now***. You may have to raise the border to gain access to the "Add in Now" text. This will load the Toolbox for use as shown in Figure 3. After selecting ***Add in Now*** you may have to configure the tool box. To configure the tool box, left click on ***Configure Now*** (not shown). A dialog box will open (also not shown). Close the dialog box.

Figure 3   Left Click Here

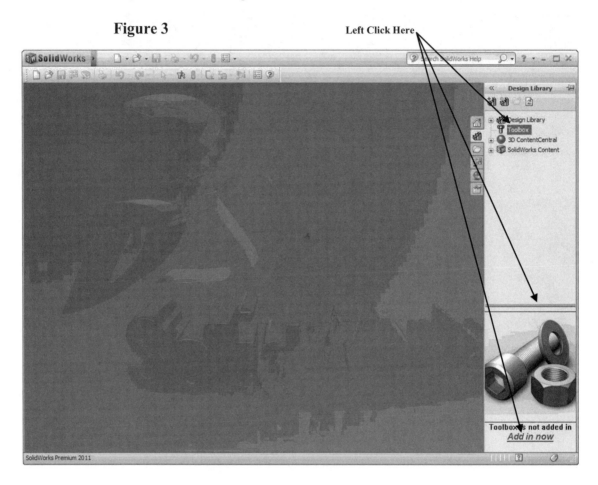

Chapter 14: Introduction to the Design Library

5. Move the cursor to the lower right portion of the screen and left click on the **Ansi Inch** folder as shown in Figure 4

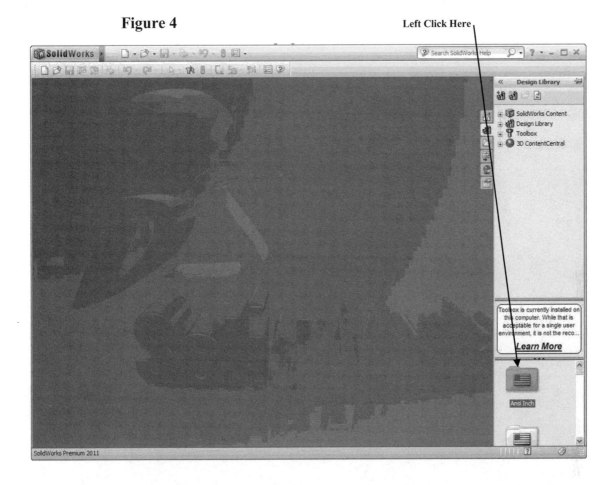

Figure 4

Left Click Here

Chapter 14: Introduction to the Design Library

6. The Toolbox will open at the right of the screen as shown in Figure 5. The upper left portion of the screen will display all the different types of items available in the Toolbox. Some items in the Toolbox will need additional editing once they are imported into a drawing file. Move the cursor to the lower right portion of the screen and left click on **Bolts and Screws** as shown in Figure 5.

Figure 5

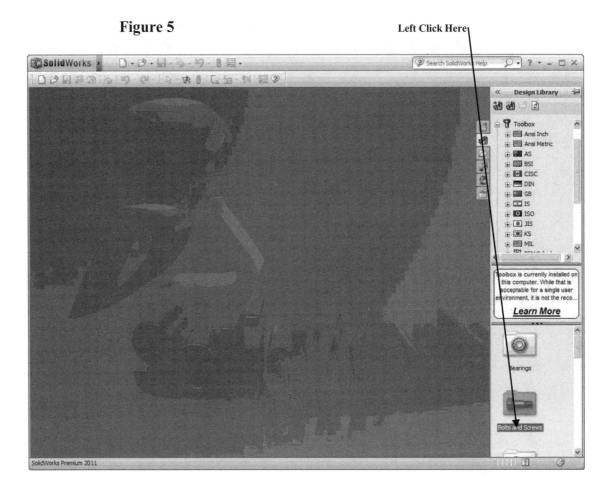

Chapter 14: Introduction to the Design Library

7. Move the cursor to the lower right portion of the screen and left click on **Hex Head** as shown in Figure 6.

Figure 6

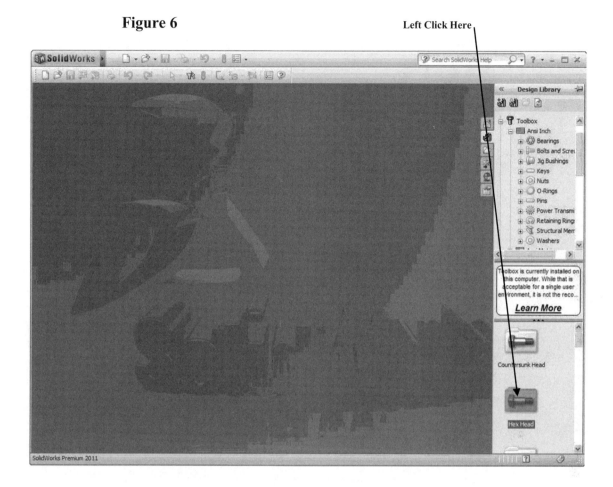

Chapter 14: Introduction to the Design Library

8. Move the cursor to the lower right portion of the screen and left click on **Heavy Hex Bolt** as shown in Figure 7.

Figure 7

9. Right click on **Heavy Hex Bolt**. A pop up menu will appear. Left click on **Create Part**... as shown in Figure **8**.

Figure 8

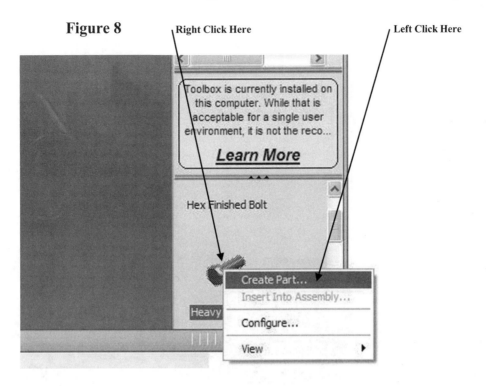

Chapter 14: Introduction to the Design Library

10. SolidWorks will create a Heavy Hex bolt. Your screen should look similar to Figure 9.

Figure 9

11. The Properties of the bolt can be edited. Move the cursor to the middle left portion of the screen and left click on the drop down arrow under "Size". Left click on ¾ -16. This refers to the diameter and pitch of the bolt. This particular bolt is ¾ (.75) inches in diameter and has 16 threads per inch. Notice that the length of the bolt will change. Threads will NOT be displayed as shown in Figure 10.

Figure 10

Chapter 14: Introduction to the Design Library

12. Move the cursor to the middle left portion of the screen and left click on the drop down arrow under "Length". Left click on **1.5**. This refers to the length of the bolt. This particular bolt is 1.5 inches in length as shown in Figure 11.

Figure 11

Chapter 14: Introduction to the Design Library

13. Left click on the green check mark as shown in Figure 12.

Figure 12

Chapter 14: Introduction to the Design Library

14. SolidWorks will calculate the properties that were entered, and generate a bolt from these properties. The bolt is ¾ inches in diameter with 16 threads per inch and 1 and ½ inches in length as shown in Figure 13. If there is a need, changes to the bolt can be accomplished in the same manner as editing all other existing solid models (Refer to Chapter 5 Editing Existing Solid Models).

Figure 13 **Left Click Here**

15. Save the bolt for use in a future Assembly drawing file.

16. Import other parts from the Toolbox for additional practice.

Chapter 15 Introduction to the DFMXpress Analysis Wizard

Objectives:
- Learn to create a simple part with the idea of designing for manufacture
- Learn to use the DFMXxpress Analysis Wizard

   Chapter 15 includes instruction on how to design and evaluate the part shown below.

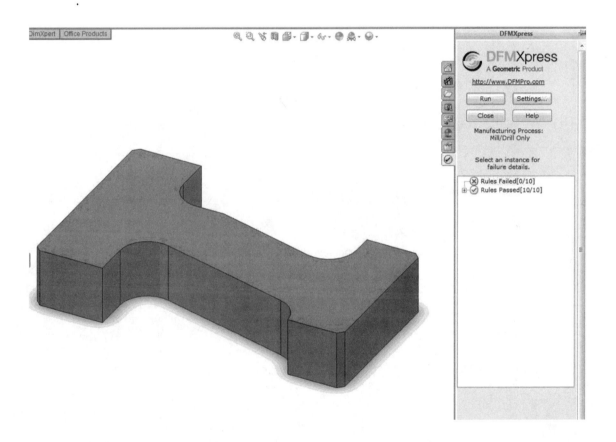

1. Complete the following sketch on the Top Plane and Extrude it to a distance of 1 inch as shown in Figure 1.

Figure 1

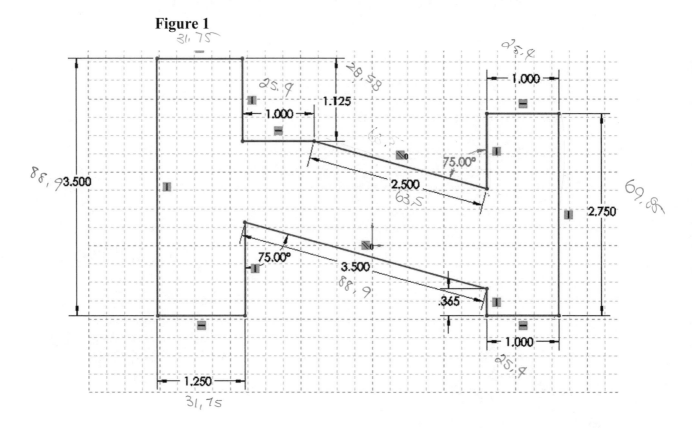

2. Use the **Fillet** command to create .5 inch radius fillets on the outside corners of the part as shown in Figure 2. Save the file where it can be retrieved later.

Figure 2

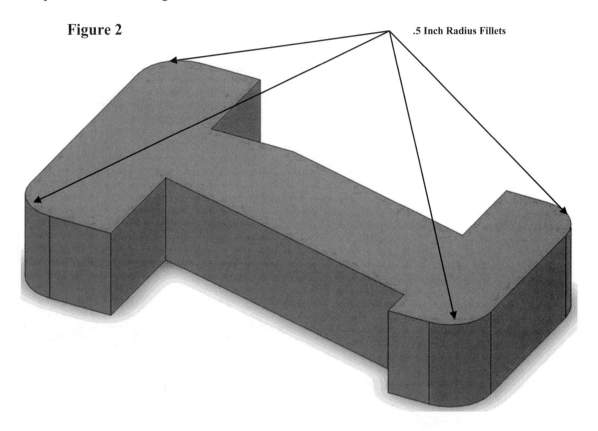

.5 Inch Radius Fillets

Chapter 15: Introduction to the DMFXpress Analysis Wizard

3. Move the cursor to the upper left portion of the screen and left click on the **Evaluate** tab as shown in Figure 3.

Figure 3   Left Click Here

4. Move the cursor to the upper right portion of the screen and left click the on the **DFMXpress Analysis Wizard** icon as shown in Figure 4.

Figure 4   Left Click Here

5. The DFMXpress Analysis Wizard will appear as shown in Figure 5.

Figure 5

518

Chapter 15: Introduction to the DMFXpress Analysis Wizard

6. Left click on **Run** as shown in Figure 6. The DFMXpress Analysis Wizard will begin evaluating the part for manufacture.

Figure 6

Left Click Here

Chapter 15: Introduction to the DMFXpress Analysis Wizard

7. Once the evaluation is complete the DFMXrpress Analysis Wizard will reveal any problems that might occur with the manufacture of the part. Left click on the "plus" signs to show a complete evaluation. The evaluation was performed for milling and drilling processes only as shown in Figure 7.

Figure 7

Chapter 15: Introduction to the DMFXpress Analysis Wizard

8. Out of 10 Rules the current design failed 2 of them. Move the cursor over one of the "Rules Failed" in the list. A detailed pop up message will appear explaining the problem and a possible solution as shown in Figure 8.

Figure 8

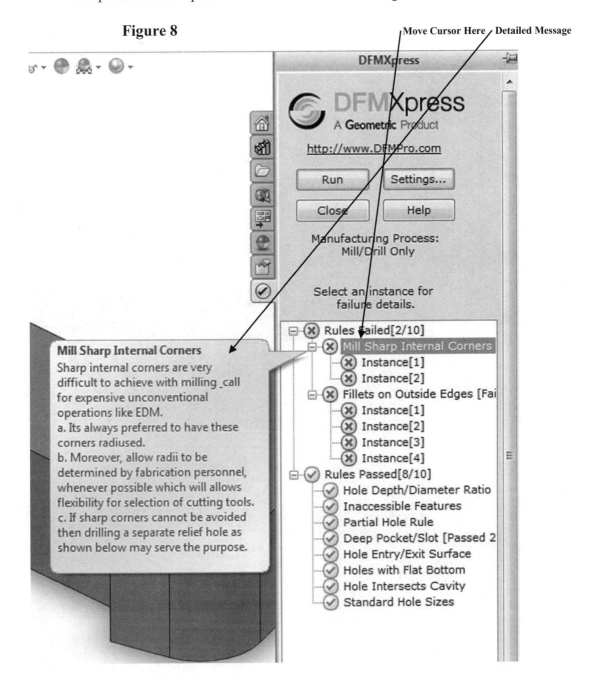

Chapter 15: Introduction to the DMFXpress Analysis Wizard

9. To address concerns the DFMXpress Analysis Wizard identified, create inside fillets with a .5 inch radius. Also delete the outside fillets and replace them with .5 equal distance chamfers (not shown) as shown in Figure 9.

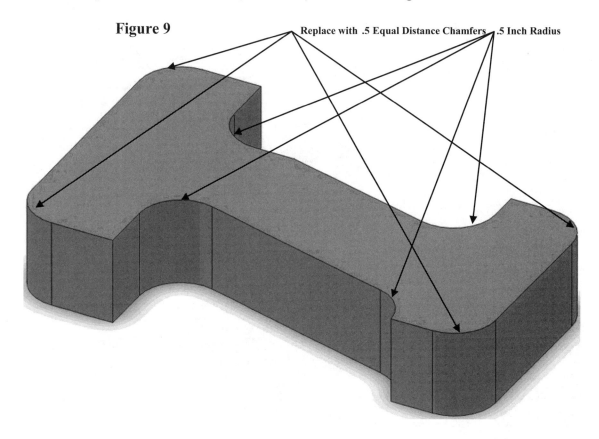

Figure 9 — Replace with .5 Equal Distance Chamfers — .5 Inch Radius

Chapter 15: Introduction to the DMFXpress Analysis Wizard

10. Left click on **Run**. All 10 Rules passed the evaluation. The DMFXpress Analysis Wizard found no concerns with the current design as shown in Figure 10.

Figure 10

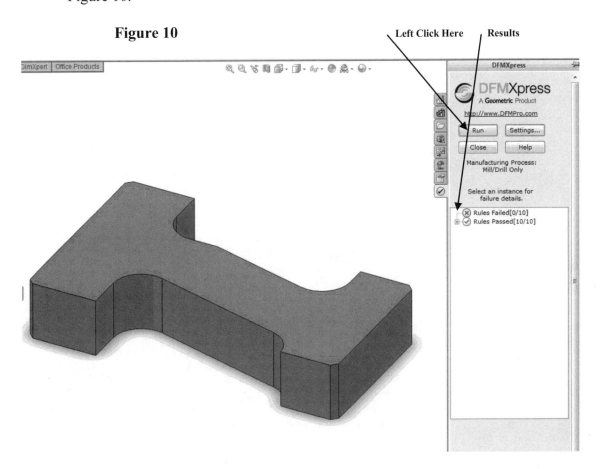

Chapter 15: Introduction to the DMFXpress Analysis Wizard

11. Feel free to create other parameters for evaluation. This is done by left clicking on **the Settings/Back button** and creating your own parameters as shown in Figure 11.

Figure 11

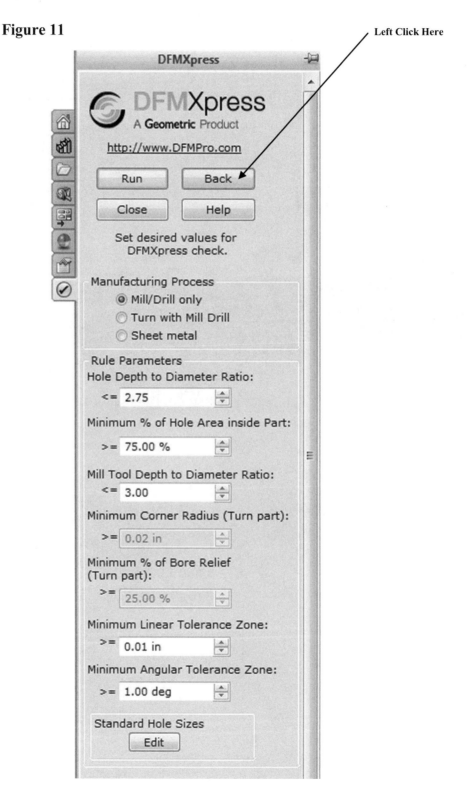

Left Click Here

Index

A

A – Landscape, 167
Annotations, 188
ANSI, 17
Assembly, 329
Auxiliary View, 182
Axis, 116, 220

B

Base Extrusion, 251
Beginning end point of a line, 10, 78
Blind (hole), 45
Browse, 329, 331

C

Cam, 448
Chamfer, 144
Chamfer1, 246
Circle, 55, 105
Circular Pattern, 118, 222
CirPattern1, 243, 249
Concentric1, 341
Confirm Delete dialog box, 186
Convert Entities, 273 - 275
Create Drawing commands, 169
Cut-Extrusion1, 240

D

Delete, 30, 110
Detailing, 17
Dimensioning standard, 17
Display Grid, 8
Display Style, 168

Distance distance, 145
Distance1, 348
Document Properties - Detailing dialog box, 8, 17
Document Properties Tab, 8, 17
Drawings, 182

E

Edit Assembly: Assem1, 367, 371
Edit Color icon, 375, 379
Edit Feature, 235, 240
Edit Sketch, 230, 375
Equal Spacing, 119
Exit Sketch, 41, 58
Extend Entities, 32, 141
Extrude, 268
Extrude dialog box, 44, 45
Extrude1, 235, 250
Extrude1 dialog box, 236
Extruded Boss/Base, 43, 141
Extruded Cut, 59, 113

F

Features commands, 42, 59
Fillet, 46, 225
Front, 210, 231
Front Plane, 5
Full Preview, 47, 145

G

Glasses icon, 45, 59
Grid/Snap, 8

H

Hidden Lines Visible icon, 168
Hide, 192
Hole Definition dialog box, 66
Hole Placement dialog box, 67
Hole Wizard command, 61, 65
Helix (spring), 477

I

Insert Component, 339, 341
Insert Sketch, 54, 62
Importing .DWG Files, 487

L

Left, 150
Line icon, 10, 23
Lofted Boss/Base, 408

M

Make Dimension Driven dialog box, 138
Make Drawing icon, 167
Mate, 340, 344,
Mechanical Mates, 447
Modify dialog box, 19, 22
Move Component, 355
Motion Study, 375

N

New Document icon, 2
New SolidWorks Document dialog box, 3, 336
Ninety degree box, 14
Normal To, 316, 319
Note, 192

O

Open dialog box, 180, 330
Open icon, 180
Options, 7, 16

P

Pan icon, 71
Parameters, 119
Part Tree, 227, 246
Plane, 399
Point, 63
Projected View, 172

R

Reference Geometry, 116, 222
Revolved Boss/Base, 94
Right, 99, 272
Right plane, 267, 269
Rotary Motor, 389
Rotate component, 347, 364
Rotate icon, 49, 50

S

Scale, 168
Section View, 186
Select, 23, 25
Shaded with Edges icon, 281
Sheet Format/Size dialog box, 167
Shell, 284
SimulationXpress, 413, Simulation, 381, 382
Sketch Commands, 54
Sketch icon, 5
Sketch1, 229
Sketcher Confirm Delete dialog box, 110
Smart Dimension, 18, 33
Snapped to the intersection, 14
Standard Views icon, 42, 51
Starting SolidWorks, 2, 78
Suppress, 251
Swept Boss/Base, 391

T

Thin Feature, 96
Tools, 7, 16
Transparent, 376
Trim Entities, 29, 31
Trimetric, 42, 52

U

Use custom scale, 168

V

View Orientation, 69

W

Wire Frame Display icon, 269
Work Plane Creation, 455

Z

Zoom icon, 83
Zoom In/Out icon, 25, 71
Zoom to Fit icon, 71, 83
Zoom Window icon, 71, 122